Basic Geometry for College Students

Alan S. Tussy
Citrus College

R. David Gustafson
Rock Valley College

THOMSON

BROOKS/COLE

Australia • Canada • Mexico • Singapore • Spain
United Kingdom • United States

THOMSON

✷

BROOKS/COLE

Editor: *Jennifer Huber*
Assistant Editor: *Rachael Sturgeon*
Editorial Assistant: *Jonathan Wegner*
Marketing Manager: *Leah Thomson*
Marketing Assistant: *Maria Salinas*
Project Manager, Editorial Production: *Ellen Brownstein*
Print/Media Buyer: *Kristine Waller*
Production Service: *Hoyt Publishing Services*

Text Design: *Vernon T. Boes* and *John Edeen*
Copy Editor: *David Hoyt*
Illustrator: *Lori Heckelman*
Cover Designer: *Roy R. Neuhaus*
Cover Image: *George Abe*
Compositor: *The Clarinda Company*
Printer: *CTPS*

Printed in China

7   06

For more information about our products, contact us at:
Thomson Learning Academic Resource Center
1-800-423-0563

For permission to use material from this text, contact us by:
Phone: 1-800-730-2214 **Fax:** 1-800-730-2215
Web: http://www.thomsonrights.com

Library of Congress Control Number: 2002108818

ISBN 0-534-39180-X

Brooks/Cole–Thomson Learning
511 Forest Lodge Road
Pacific Grove, CA 93950
USA

Asia
Thomson Learning
5 Shenton Way #01-01
UIC Building
Singapore 068808

Australia
Nelson Thomson Learning
102 Dodds Street
South Melbourne, Victoria 3205
Australia

Canada
Nelson Thomson Learning
1120 Birchmount Road
Toronto, Ontario M1K 5G4
Canada

Europe/Middle East/ Africa
Thomson Learning
High Holborn House
50/51 Bedford Row
London WC1R 4LR
United Kingdom

Latin America
Thomson Learning
Seneca, 53
Colonia Polanco
11560 Mexico D.F.
Mexico

Spain
Paraninfo Thomson Learning
Calle/Magallanes, 25
28015 Madrid, Spain

CONTENTS

Contents

PREFACE

For the Instructor

Geometry may be the area of study that is the most overlooked in today's college mathematics curriculum. This is unfortunate; the study of geometry is very beneficial. It develops students' logic and reasoning skills and improves their ability to spot valid and invalid arguments. When studying geometry, students visualize in two and three dimensions, and they gain a better understanding of measurement and units. Most important, students have an opportunity to couple their algebra skills with the geometry concepts that they study to solve meaningful application problems.

Basic Geometry for College Students has been written to address the need for a concise overview of fundamental geometry topics. Sections 1–7, the first part of the text, introduce such topics as angles, polygons, perimeter, area, and circles. In the second part of the text, Sections 8–11 cover congruent and similar triangles, special triangles, volume, and surface area. Appendix I explains inductive and deductive reasoning.

This text can be used for a short introductory course (5–6 weeks) in geometry. It can also serve as a geometry supplement for an Elementary Algebra course, an Intermediate Algebra course, or a combination Elementary/Intermediate Algebra course.

Basic Geometry for College Students uses a variety of instructional approaches that reflect the recommendations of NCTM and AMATYC. You will find the vocabulary, practice, and well-defined pedagogy of a traditional appproach. We also emphasize the reasoning, modeling, communicating, and technological skills that are such a big part of the current reform movement.

Acknowledgments

The authors wish to express their gratitude to the Mathematics Department of Century College, White Bear Lake, Minnesota, whose innovative curriculum revision provided the impetus for this project. We offer special thanks to Professor Beth Hentges and Professor Carol Purcell for their helpful input and diligent proofreading.

Without the talents and dedication of the editorial, production, and sales staff of Brooks/Cole, this text would not have been so well accomplished. We express our sincere appreciation of the hard work of Jennifer Huber, Michelle Clayton, and Ellen Brownstein, as well as the freelance talents of David Hoyt and Lori Heckelman and the superb typesetting of the Clarinda Company.

Alan S. Tussy
R. David Gustafson

*G*EOMETRY COMES FROM THE *G*REEK WORDS GEO (*MEANING EARTH*) AND METRON (*MEANING MEASURE*).

1 *Basic Geometric Figures*

In this section, you will learn about
- Points, lines, and planes • Line segments and rays • Angles
- Measuring angles

INTRODUCTION. Geometry is a branch of mathematics that studies the properties of two- and three-dimensional figures such as triangles, circles, cylinders, and spheres. More than 5,000 years ago, Egyptian surveyors used geometry to measure areas of land in the flooded plains of the Nile River after heavy spring rains. Even today, engineers marvel at the Egyptians' use of geometry in the design and construction of the pyramids. History records many other practical applications of geometry made by Babylonians, Chinese, Indians, and Romans.

The classical Greeks (660–300 B.C.E.) are credited with refining years of practical use of geometry into a systematic subject of logical thought. **Pythagoras** (580?–501? B.C.E.) is generally regarded as the first of the great Greek mathematicians. Although very little is known about him personally, there is much fascinating literature about the Order of Pythagoreans—a private academic society that he established. Pythagoras and his students devoted themselves to the study of mathematics, astronomy, and philosophy, and they developed geometry into an abstract science.

Many scholars consider **Euclid** (330?–275? B.C.E.) to be the greatest of the Greek mathematicians. His book, *The Elements,* is an impressive study of geometry and number theory. It presents geometry in a highly structured form that begins with several simple assumptions and then expands on them using logical reasoning. For more than 2,000 years, *The Elements* was the textbook that students all over the world used to learn geometry.

Points, lines, and planes

Geometry is based on three undefined words: **point, line,** and **plane.** Although we will make no attempt to define these words formally, we can think of a point as a geometric figure that has position but no length, width, or depth. Points are labeled with capital letters. For example, point *A* is shown in Figure 1(a).

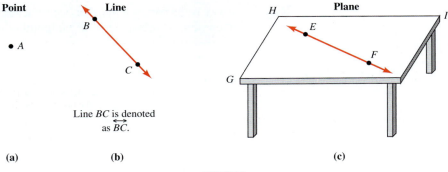

Line *BC* is denoted as $\overleftrightarrow{BC}$.

(a)   (b)   (c)

FIGURE 1

A line is infinitely long but has no width or depth. In Figure 1(b), the line that passes through points *B* and *C* is denoted as $\overleftrightarrow{BC}$.

A plane is a flat surface, like a table top, that has length and width but no depth. In Figure 1(c), $\overleftrightarrow{EF}$ lies in plane *GHI*.

As Figure 1(b) illustrates, points *B* and *C* determine exactly one line, the line $\overleftrightarrow{BC}$. In Figure 1(c), the points *E* and *F* determine exactly one line, the line $\overleftrightarrow{EF}$. In general, any two different points determine exactly one line.

Other geometric figures can be created by using parts or combinations of points, lines, and planes.

Line segments and rays

Line segment

> The **line segment** *AB*, denoted as $\overline{AB}$, is the part of a line that consists of points *A* and *B* and all points in between (see Figure 2). Points *A* and *B* are the **endpoints** of the segment.

Line segment *AB*
is denoted
as $\overline{AB}$.

FIGURE 2

Every line segment has a **midpoint,** which divides the segment into two parts of equal length. In Figure 3, *M* is the midpoint of segment *AB*, because the measure of $\overline{AM}$ (denoted as m($\overline{AM}$)) is equal to the measure of $\overline{MB}$ (denoted as m($\overline{MB}$)).

$$m(\overline{AM}) = 4 - 1$$
$$= 3$$

and

$$m(\overline{MB}) = 7 - 4$$
$$= 3$$

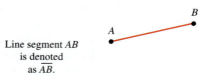

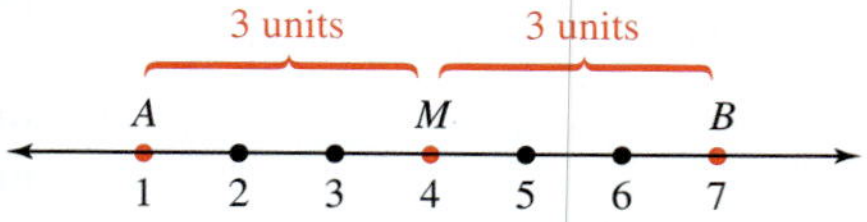

FIGURE 3

Since the measure of both segments is 3 units, m($\overline{AM}$) = m($\overline{MB}$).

When two line segments have the same measure, we say that they are **congruent.** Since m($\overline{AM}$) = m($\overline{MB}$), we can write

$$\overline{AM} \cong \overline{MB} \quad \text{Read} \cong \text{as “is congruent to.”}$$

Another geometric figure is the *ray,* as shown in Figure 4.

Ray

> A **ray** is the part of a line that begins at some point (say, *A*) and continues forever in one direction. Point *A* is the **endpoint** of the ray.

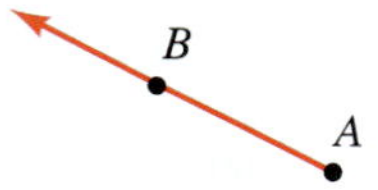

Ray *AB* is denoted as $\overrightarrow{AB}$. The endpoint
of the ray is always listed first.

FIGURE 4

COMMENT   To name a ray, we list its endpoint and then one other point on the ray. Sometimes it is possible to name a ray in more than one way. For example, in Figure 5, $\overrightarrow{DE}$ and $\overrightarrow{DF}$ name the same ray. This is because both have point D as their endpoint and extend forever in the same direction. In contrast, $\overrightarrow{DE}$ and $\overrightarrow{ED}$ are not the same ray. They have different endpoints and point in opposite directions.

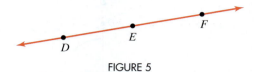

FIGURE 5

Angles

Angle

> An **angle** is a figure formed by two rays with a common endpoint. The common endpoint is called the **vertex,** and the rays are called **sides.**

The angle in Figure 6 can be denoted as

$\angle BAC$, $\angle CAB$, $\angle A$,   or $\angle 1$ The symbol $\angle$ means angle.

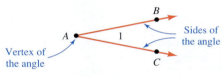

FIGURE 6

COMMENT   When using three letters to name an angle, be sure the letter name of the vertex is the middle letter. Furthermore, we can only name an angle using a single vertex letter when there is no possibility of confusion. For example, in Figure 7, we cannot refer to any of the angles as simply $\angle X$, because we would not know if that meant $\angle WXY$, $\angle WXZ$, or $\angle YXZ$.

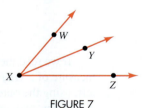

FIGURE 7

Measuring angles

One unit of measurement of an angle is the **degree.** The symbol for degree is a small raised circle, °. An angle measure of 1° (read as "one degree") means that one side of an angle is rotated $\frac{1}{360}$ of a complete revolution about the vertex from the other side of the angle. In Figure 8, the measure of $\angle ABC$ is 1°.

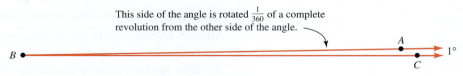

FIGURE 8

Figure 9 shows the measures of several other angles. An angle measure of 90° is equivalent to $\frac{90}{360} = \frac{1}{4}$ of a complete revolution. An angle measure of 180° is equivalent to $\frac{180}{360} = \frac{1}{2}$ of a complete revolution, and an angle measure of 270° is equivalent to $\frac{270}{360} = \frac{3}{4}$ of a complete revolution.

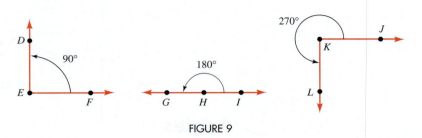

FIGURE 9

We can use a **protractor** to measure angles. Begin by placing the center of the protractor at the vertex of the angle, with the edge of the protractor aligned with one side of the angle. See Figure 10. The angle measure is found by determining where the other side of the angle crosses the scale. Be careful to use the appropriate scale, inner or outer, when reading an angle measure.

Angle	Measure in degrees
$\angle ABC$	30°
$\angle ABD$	60°
$\angle ABE$	110°
$\angle ABF$	150°
$\angle ABG$	180°

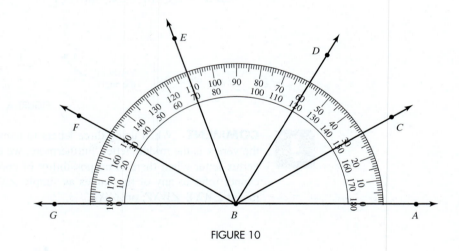

FIGURE 10

If we read the protractor from left to right, using the inner scale, we can see that the measure of $\angle GBF$ (denoted as m($\angle GBF$)) is 30°. If we read the protractor from right to left, using the outer scale, we see that m($\angle ABC$) = 30°.

When two angles have the same measure, we say that they are **congruent.** Since m($\angle ABC$) = 30° and m($\angle GBF$) = 30°, we can write

$$\angle ABC \cong \angle GBF$$

We classify angles according to their measure, as in Figure 11 on the next page.

Classification of angles

Acute angles: Angles whose measures are greater than 0° but less than 90°.

Right angles: Angles whose measures are 90°.

Obtuse angles: Angles whose measures are greater than 90° but less than 180°.

Straight angles: Angles whose measures are 180°.

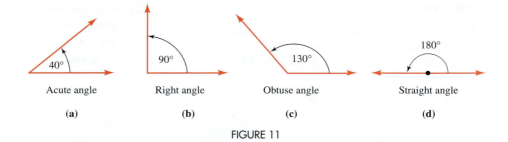

FIGURE 11

COMMENT   A ⌐ symbol is often used in diagrams to denote a right angle. For example, in Figure 12, the ⌐ symbol drawn near the vertex of ∠ABC indicates that m(∠ABC) = 90°.

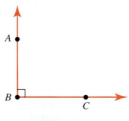

FIGURE 12

EXAMPLE 1   *Classifying angles.*   Classify each angle in Figure 13 as an acute angle, a right angle, an obtuse angle, or a straight angle.

Solution   Since m(∠1) < 90°, it is an acute angle.

Since m(∠2) > 90° but less than 180°, it is an obtuse angle.

Since m(∠BDE) = 90°, it is a right angle.

Since m(∠ABC) = 180°, it is a straight angle.

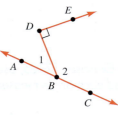

FIGURE 13   ■

STUDY SET   Section 1

VOCABULARY   *Fill in the blanks.*

1. Three undefined words in geometry are _____, _____, and _____.

2. A line _____ has two endpoints.

3. A _____ divides a line segment into two parts of equal length.

4. A _____ is the part of a line that begins at some point and continues forever in one direction.

5. An _____ is formed by two rays with a common endpoint.

6. An angle is measured in _____.

7. A _____ is used to measure angles.

8. An _____ angle is less than 90°.

9. A _____ angle measures 90°.

10. An _____ angle is greater than 90° but less than 180°.

11. The measure of a straight angle is _____.

12. When two segments have the same length, we say that they are _____. When two angles have the same measure, we say that they are _____.

CONCEPTS

13. Draw each geometric figure and label it completely.
   a. Point *T*
   b. $\overline{RS}$
   c. $\overleftrightarrow{JK}$
   d. $\overrightarrow{PQ}$
   e. ∠*XYZ*
   f. ∠*L*

14. a. Given two points (say, *M* and *N*), how many different lines pass through these two points?
   b. Fill in the blank: In general, two different points determine one _____ .

15. Give four ways to name the angle shown in Illustration 1.

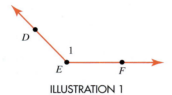

ILLUSTRATION 1

16. An angle is shown in Illustration 2.
   a. What two rays are the sides of the angle?
   b. What point is the vertex of the angle?
   c. Name the angle in three ways.

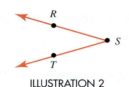

ILLUSTRATION 2

In Exercises 17–26, use the protractor in Illustration 3 to find each angle measure.

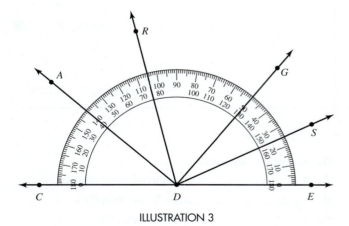

ILLUSTRATION 3

17. m(∠*GDE*)   **18.** m(∠*ADE*)
19. m(∠*EDS*)   **20.** m(∠*EDR*)
21. m(∠*CDA*)   **22.** m(∠*CDR*)
23. m(∠*CDG*)   **24.** m(∠*CDS*)
25. m(∠*CDE*)   **26.** m(∠*EDC*)

27. Estimate the measure of each angle. Do not use a protractor.

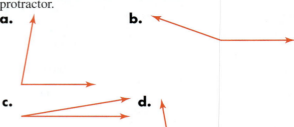

28. Draw an example of each type of angle.
   a. An acute angle   **b.** An obtuse angle
   c. A right angle   **d.** A straight angle

29. Fill in the blanks.
   a. If m($\overline{AB}$) = m($\overline{CD}$), then $\overline{AB}$ _____ $\overline{CD}$.
   b. If ∠*ABC* ≅ ∠*DEF,* then m(∠*ABC*) _____ m(∠*DEF*).

30. Refer to Illustration 4.
   a. Name $\overrightarrow{NM}$ in another way.
   b. Do $\overrightarrow{MN}$ and $\overrightarrow{NM}$ name the same ray?

ILLUSTRATION 4

NOTATION   *Fill in the blanks.*

31. The symbol ∠ means _____ .
32. The symbol $\overline{AB}$ is read as "_____ *AB*." The symbol $\overrightarrow{AB}$ is read as "_____ *AB*."
33. The symbol _____ is read as "is congruent to."
34. The symbol ⌐ indicates a _____ angle.

PRACTICE   *Refer to Illustration 5 and find the length of each segment.*

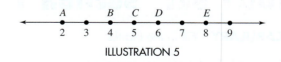

ILLUSTRATION 5

35. $\overline{AC}$ **36.** $\overline{BE}$
37. $\overline{CE}$ **38.** $\overline{BD}$
39. $\overline{CD}$ **40.** $\overline{DE}$

Refer to Illustration 5 and find each midpoint.

41. Find the midpoint of $\overline{AD}$.
42. Find the midpoint of $\overline{BE}$.

Use a protractor to measure each angle.

43.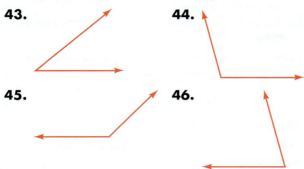

44.

45.

46.

Refer to Illustration 6 and tell whether each statement is true. If a statement is false, explain why.

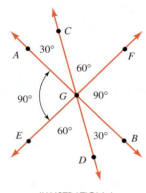

ILLUSTRATION 6

47. $\overrightarrow{GF}$ has point *G* as its endpoint.

48. $\overline{AG}$ has no endpoints.

49. Line *CD* has three endpoints.

50. Point *D* is the vertex of ∠*DGB*.

51. m(∠*AGC*) = m(∠*BGD*)

52. ∠*AGF* ≅ ∠*BGE*

53. ∠*FGB* ≅ ∠*EGA*

54. ∠*AGC* and ∠*CGF* are adjacent angles.

Refer to Illustration 6 and tell whether each angle is an acute angle, a right angle, an obtuse angle, or a straight angle.

55. ∠*AGC*   **56.** ∠*EGA*

57. ∠*FGD*   **58.** ∠*BGA*

59. ∠*BGE*   **60.** ∠*AGD*

61. ∠*DGC*   **62.** ∠*DGB*

APPLICATIONS

63. BASEBALL   Use the following definition to draw the strike zone for the player shown in Illustration 7. The strike zone is that area over home plate the upper

limit of which is a horizontal line at the midpoint between the top of the shoulders and the top of the uniform pants and the lower level is a line at the hollow beneath the kneecap.

ILLUSTRATION 7

64. PHYSICS   Illustration 8 shows a 15-pound block that is suspended with two ropes, one of which is horizontal. Classify each numbered angle in the illustration as either acute, obtuse, or right.

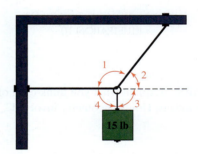

ILLUSTRATION 8

65. MUSICAL INSTRUMENTS   Suppose that you are a beginning band teacher describing the correct posture needed to play various instruments. Use the diagrams in Illustration 9 to approximate the angle measure at which each instrument should be held in relation to the student's body: **a.** flute   **b.** clarinet   **c.** trumpet

a.   **b.**   **c.**

ILLUSTRATION 9

66. PLANETS   Illustration 10 shows the direction of rotation of several planets in our solar system. It also shows the angle of tilt of each planet.
  a. Which planets have an angle of tilt that is an acute angle?
  b. Which planets have an angle of tilt that is an obtuse angle?

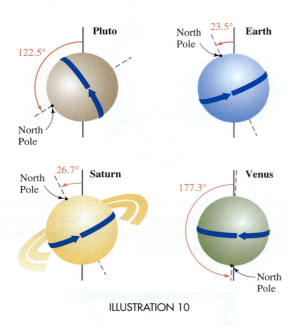

ILLUSTRATION 10

WRITING

67. PHRASES   Explain what you think each of these phrases means. How is geometry involved?

  a. The president did a complete 180-degree flip on the subject of a tax cut.
  b. The rollerblader did a "360" as she jumped off the ramp.

68. In the statements below, the ° symbol is used in two different ways. Explain the difference.

$$85°F \qquad \text{and} \qquad m(\angle A) = 85°$$

69. What is a protractor?

70. Explain the difference between a ray and a line segment.

71. Explain why we cannot refer to any of the angles in Illustration 11 as $\angle T$.

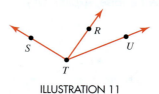

ILLUSTRATION 11

72. Describe what is wrong with the drawing of $\angle R$ in Illustration 12.

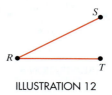

ILLUSTRATION 12

2 *More about Angles*

In this section, you will learn about

- Adjacent angles • Vertical angles
- Complementary and supplementary angles

INTRODUCTION. Figure 14 shows three geometric drawings where rays and lines intersect to form several angles. Often, a pair (or pairs) of angles in diagrams like these are related. In this section, we will explore several important relationships that exist between certain pairs of angles. We will also use some of the algebraic skills that we have studied to find unknown angle measures in the drawings shown below.

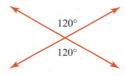
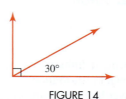
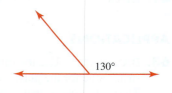

FIGURE 14

Adjacent angles

Two angles that have a common vertex and a common side are called **adjacent angles** if they are side-by-side and their interiors do not overlap.

EXAMPLE 1   *Evaluating angles.*   Two angles with degree measures of x and 35° are adjacent angles. Use the information in Figure 15 to find x.

Solution

We can use algebra to find the unknown angle measure labeled x. Since the sum of the measures of the angles is 80°, we have

$$x + 35° = 80°$$

$x + 35° - 35° = 80° - 35°$ To undo the addition of 35°, subtract 35° from both sides.

$x = 45°$ Do the subtractions: $35° - 35° = 0°$ and $80° - 35° = 45°$.

Thus, $x = 45°$.

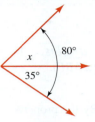

FIGURE 15

Self Check

In the figure below, find x.

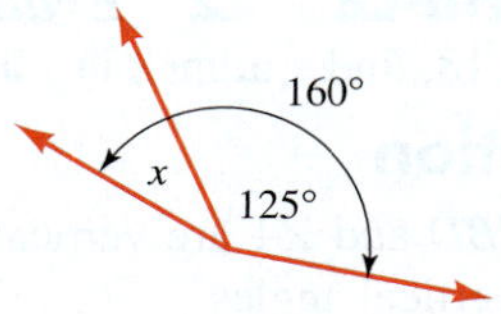

Answer: 35°

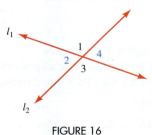

COMMENT   In Figure 15, we used the variable x to represent an unknown angle measure. In such cases, we will assume that the variable "carries" with it the associated units of degrees. That means we do not have to write a ° symbol next to the variable. Furthermore, if x represents an unknown number of degrees, then expressions such as $3x, x + 15°$, and $4x - 20°$ also have units of degrees.

Vertical angles

When two lines intersect, pairs of nonadjacent angles are called **vertical angles.** In Figure 16, lines l_1 (read as "line l sub 1") and l_2 (read as "line l sub 2") intersect. $\angle 1$ and $\angle 3$ are vertical angles, as are $\angle 2$ and $\angle 4$.

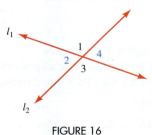

FIGURE 16

To illustrate that vertical angles always have the same measure, we refer to Figure 17, with angles having measures of x, y, and 30°. Since the measure of any straight angle is 180°, we have

$$30° + x = 180° \quad \text{and} \quad 30° + y = 180°$$
$$x = 150° \qquad\qquad y = 150°$$

To undo the addition of 30°, subtract 30° from both sides.

Since x and y are both 150°, $x = y$.

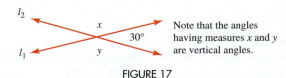

Note that the angles having measures x and y are vertical angles.

FIGURE 17

The previous example illustrated that vertical angles have the same measure. When two angles have the same measure, we say that they are *congruent*. Therefore, we have the following important fact.

Property of vertical angles	Vertical angles are congruent (have the same measure).

EXAMPLE 2 *Evaluating angles.* In
Figure 18, find **a.** m($\angle 1$) and **b.** m($\angle ABF$).

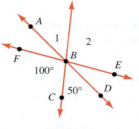

FIGURE 18

Solution

a. $\angle CBD$ and $\angle 1$ are vertical angles. By the property of vertical angles,

$\angle CBD \cong \angle 1$ Read as "Angle *CBD* is congruent to angle one."

Since congruent angles have the same measure,

m($\angle CBD$) = m($\angle 1$)

In the figure, we are given m($\angle CBD$) = 50°. Thus, m($\angle 1$) is also 50°, and we can write m($\angle 1$) = 50°.

b. Since $\angle ABD$ is a straight angle, the sum of the measures of $\angle ABF$, the 100° angle, and the 50° angle is 180°. If m($\angle ABF$) = x, we have

$x + 100° + 50° = 180°$

$\qquad x + 150° = 180°$ Do the addition: 100° + 50° = 150°.

$\qquad\qquad x = 30°$ Subtract 150° from both sides: 180° − 150° = 30°.

Thus, m($\angle ABF$) = 30°

Self Check

In Figure 18, find

a. m($\angle 2$)

b. m($\angle DBE$)

Answers: a. 100°, **b.** 30° ■

EXAMPLE 3 *Evaluating angles.* In
Figure 19, find x. Then determine m($\angle ABC$) and m($\angle CBE$).

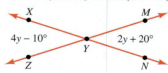

FIGURE 19

Solution

In the figure, two vertical angles have degree measures that are represented by the algebraic expressions $4x - 20°$ and $3x + 15°$. Since the angles are vertical angles, they have equal measures.

$\qquad 4x - 20° = 3x + 15°$

$4x - 20° \mathbf{- 3x} = 3x + 15° \mathbf{- 3x}$ To eliminate 3x from the right-hand side, subtract 3x from both sides.

$\qquad\quad x - 20° = 15°$ Combine like terms: 4x − 3x = x and 3x − 3x = 0.

$\qquad\qquad\quad x = 35°$ To undo the subtraction of 20°, add 20° to both sides.

Thus, x = 35°.

To find m($\angle ABC$), we evaluate the expression $3x + 15°$ for x = 35°.

$3x + 15° = 3(35°) + 15°$ Substitute 35° for x.

$\qquad\quad = 105° + 15°$ Do the multiplication.

$\qquad\quad = 120°$

Thus, m($\angle ABC$) = 120°.

$\angle ABE$ is a straight angle. Since the measure of a straight angle is 180° and m($\angle ABC$) = 120°, m($\angle CBE$) must be 180° − 120° or 60°.

Self Check

In the figure below, find y. Then determine m($\angle XYZ$) and m($\angle MYX$).

Answer: 15°, 50°, 130° ■

EXAMPLE 4   *Using algebra to solve a geometry problem.*   The measure of one angle is 5 degrees more than a number, and an angle vertical to it measures 25° less than six times the number. What is the measure of the first angle?

Analyze the problem

- There are two angles to consider, and they are vertical angles.
- The measure of one angle is 5° more than a number.
- The measure of the other angle is 25° less than six times the same number.
- We are to find the measure of the first angle.

Form an equation   Since the measure of each angle is expressed in terms of the same number, we begin by letting n = the number. When we translate the key phrases *5° more than the number* and *25° less than six times the number,* we see that each angle measure can be represented by an algebraic expression.

$n + 5°$ = the degree measure of the first angle

$6n - 25°$ = the degree measure of the second angle

At this stage of the solution process, a sketch of the situation is often helpful. See Figure 20.

FIGURE 20

Because the angles are vertical angles, they are congruent, and we have

The measure of the first angle	is equal to	the measure of the second angle.
$n + 5°$	=	$6n - 25°$

Solve the equation

$$n + 5° = 6n - 25°$$

$n + 5° - \boldsymbol{n} = 6n - 25° - \boldsymbol{n}$	To eliminate n on the left-hand side, subtract n from both sides.
$5° = 5n - 25°$	Combine like terms: $n - n = 0$ and $6n - n = 5n$.
$5° + \boldsymbol{25°} = 5n - 25° + \boldsymbol{25°}$	To undo the subtraction of 25°, add 25° to both sides.
$30° = 5n$	Do the additions.
$\dfrac{30°}{\boldsymbol{5}} = \dfrac{5n}{\boldsymbol{5}}$	To undo the multiplication by 5, divide both sides by 5.
$6° = n$	Do the divisions.

The number is 6°. To find the measure of the first angle, we evaluate the expression $n + 5°$ for $n = 6°$.

$$\boldsymbol{n} + 5° = \boldsymbol{6°} + 5° \quad \text{Substitute 6° for } n.$$
$$= 11°$$

State the conclusion   The measure of the first angle is 11°.

Check the result   If we evaluate $6\boldsymbol{n} - 2°$ for $n = 6°$ to find the measure of the other angle, we get $6(\boldsymbol{6°}) - 25° = 11°$. The measures of the vertical angles are the same, 11°. The result of $n = 6°$ checks.   ■

Complementary and supplementary angles

Complementary and supplementary angles

Two angles are **complementary angles** when the sum of their measures is 90°.
Two angles are **supplementary angles** when the sum of their measures is 180°.

In Figure 21(a), ∠ABC and ∠CBD are complementary angles because the sum of their measures is 90°. Each angle is said to be the **complement** of the other. In Figure 21(b), ∠X and ∠Y are also complementary angles, because m(∠X) + m(∠Y) = 90°. This figure illustrates an important fact: Complementary angles need not be adjacent angles.

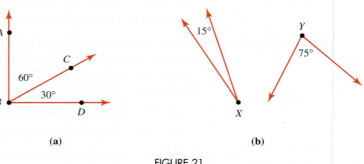

(a)   (b)

FIGURE 21

In Figure 22(a), ∠MNO and ∠ONP are supplementary angles, because the sum of their measures is 180°. Each angle is said to be the **supplement** of the other. Supplementary angles need not be adjacent angles. For example, in Figure 22(b), ∠G and ∠H are supplementary angles, because m(∠G) + m(∠H) = 180°.

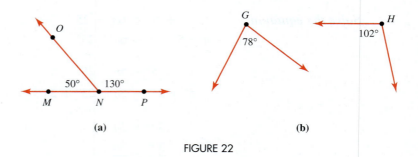

(a)   (b)

FIGURE 22

 COMMENT   The definition of supplementary angles requires that the sum of *two* angles be 180°. Three angles of 40°, 60°, and 80° are not supplementary even though their sum is 180°.

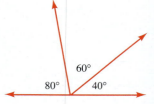

EXAMPLE 5   *Finding the complement and supplement of an angle.*
a. Find the complement of a 35° angle.
b. Find the supplement of a 105° angle.

Solution
a. See Figure 23. Let *x* represent the measure of the complement of the 35° angle. Since the angles are complementary, we have

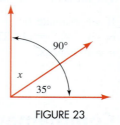

FIGURE 23

$$x + 35° = 90°$$ The sum of the angles' measures must be 90°.

$$x = 55°$$ To undo the addition of 35°, subtract 35° from both sides: 90° − 35° = 55°.

The complement of a 35° angle has measure 55°.

Self Check
a. Find the complement of a 50° angle.

b. Find the supplement of a 50° angle.

b. See Figure 24. Let *y* represent the measure of the supplement of the 105° angle. Since the angles are supplementary, we have

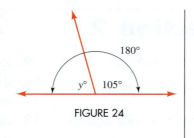

$$y + 105° = 180°$$ The sum of the angles' measures must be 180°.

$$y = 75°$$ To undo the addition of 105°, subtract 105° from both sides: $180° - 105° = 75°$.

The supplement of a 105° angle has measure 75°.

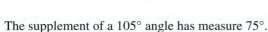

FIGURE 24

Answers: a. 40°,   **b.** 130°   ■

EXAMPLE 6   *Using algebra to solve a geometry problem.*   The measure of one angle is 15° more than twice the measure of its supplement. What is the measure of the angle?

Analyze the problem
- There are two angles to consider.
- The angles are supplementary.
- The measure of one angle (the first angle) is 15° more than twice the measure of the other angle (its supplement).
- We are to find the measure of the first angle.

Form an equation
Since the measure of one of the angles is expressed in terms of its supplement, we let $x =$ the measure of the supplement. Translating the given key words *more than* and *twice,* we can represent the measure of the first angle by the expression

$$2x + 15°$$

Because the angles are supplements, we have

The measure of the first angle	plus	the measure of its supplement	is	180°.
$2x + 15°$	$+$	x	$=$	$180°$

Solve the equation

$$2x + 15° + x = 180°$$
$$3x + 15° = 180°$$ Combine like terms: $2x + x = 3x$.
$$3x + 15° - 15° = 180° - 15°$$ To undo the addition of 15°, subtract 15° from both sides.
$$3x = 165°$$ Do the subtractions
$$\frac{3x}{3} = \frac{165°}{3}$$ To undo the multiplication by 3, divide both sides by 3.
$$x = 55°$$ Do the divisions.

The measure of the supplement is 55°. To find the measure of the first angle, we evaluate the expression $2x + 15°$ for $x = 55°$.

$$2x + 15° = 2(55°) + 15°$$ Substitute 55° for *x*.
$$= 110° + 15°$$ Do the multiplication.
$$= 125°$$

State the conclusion   The measure of the first angle is 125°.

Check the result   Since 125° is 15° more than twice 55°, and since $125° + 55° = 180°$, the result of 125° checks.   ■

STUDY SET Section 2

VOCABULARY   *Fill in the blanks.*

1. _____ angles have the same vertex and are side-by-side, and their interiors do not overlap.

2. When two lines intersect, pairs of nonadjacent angles are called _____ angles.

3. The sum of two _____ angles is 180°.

4. The sum of two complementary angles is _____ .

5. When two angles have the same measure, we say that they are _____ .

6. The word *sum* indicates the operation of _____ .

CONCEPTS

7. Fill in the blanks.

   a. If $\angle MNO \cong \angle BFG$, then m($\angle MNO$) _____ m($\angle BFG$).

   b. Vertical angles are _____ .

   c. An angle with degree measure 33 more than a number can be represented by the expression $n +$.

   c. An angle with degree measure 5 less than twice a number can be represented by the expression $- 5°$.

8. Refer to Illustration 1. Fill in the blanks.

   a. $\angle XYZ$ and $\angle$ _____ are vertical angles.

   b. $\angle XYZ$ and $\angle ZYW$ are _____ angles.

   c. $\angle ZYW$ and $\angle XYV$ are _____ angles.

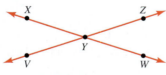

ILLUSTRATION 1

Refer to Illustration 2 and tell whether each statement is true.

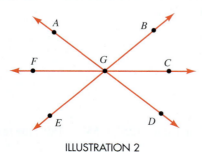

ILLUSTRATION 2

9. $\angle AGF$ and $\angle DGC$ are vertical angles.

10. $\angle FGE$ and $\angle BGA$ are vertical angles.

11. m($\angle AGB$) = m($\angle BGC$).

12. $\angle AGC \cong \angle DGF$.

Refer to Illustration 3 and tell whether each pair of angles are congruent.

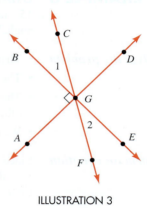

ILLUSTRATION 3

13. $\angle 1$ and $\angle 2$

14. $\angle FGB$ and $\angle CGE$

15. $\angle AGB$ and $\angle DGE$

16. $\angle CGD$ and $\angle CGB$

17. $\angle AGF$ and $\angle FGE$

18. $\angle AGB$ and $\angle BGD$

Refer to Illustration 3 and tell whether each statement is true.

19. $\angle 1$ and $\angle CGD$ are adjacent angles.

20. $\angle 2$ and $\angle 1$ are adjacent angles.

21. $\angle FGA$ and $\angle AGC$ are supplementary.

22. $\angle AGB$ and $\angle BGC$ are complementary.

23. $\angle AGF$ and $\angle 2$ are complementary.

24. $\angle AGB$ and $\angle EGD$ are supplementary.

25. $\angle EGD$ and $\angle DGB$ are supplementary.

26. $\angle DGC$ and $\angle AGF$ are complementary.

NOTATION   *Fill in the blanks.*

27. The symbol $\angle$ means _____ .

28. The symbol $\cong$ is read as "is _____ to."

29. A _____ is a letter, such as *x*, that is used to stand for a number.

30. The symbol l_1 can be used to name a line. It is read as "line *l* _____ 1."

PRACTICE   *Refer to Illustration 4, in which m(∠1) = 50°. Find the measure of each angle or sum of angles.*

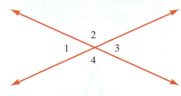

ILLUSTRATION 4

31. ∠4   **32.** ∠3

33. m(∠1) + m(∠2) + m(∠3)

34. m(∠2) + m(∠4)

Refer to Illustration 5, in which m(∠1) + m(∠3) + m(∠4) = 180°, ∠3 ≅ ∠4, and ∠4 ≅ ∠5. Find the measure of each angle.

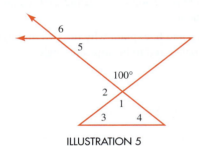

ILLUSTRATION 5

35. ∠1   **36.** ∠2

37. ∠3   **38.** ∠6

Find x.

39.   **40.**

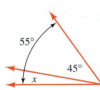   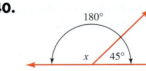

41.   **42.**

In Exercises 43–44, first find x. Then find m(∠ABD) and m(∠DBE).

43.   **44.**

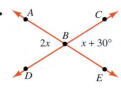   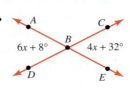

In Exercises 45–46, first find x. Then find m(∠ZYQ) and m(∠PYQ)

45.   **46.**

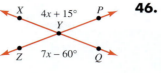   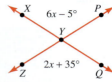

Write an equation to solve each problem.

47. Two angles are vertical angles. The first angle has a measure that is 50° more than a number. The other angle has a measure that is 23° more than ten times the number. Find the measure of the first angle.

48. Two angles are vertical angles. The first angle has a degree measure that is three times a number. The other angle has a measure that is 30° less than five times the number. Find the measure of the first angle.

49. The measure of one angle is 15° more than twice a number, and an angle vertical to it measures 5° less than four times the number. What is the measure of the first angle?

50. The degree measure of one angle is half of a number, and an angle vertical to it measures 15° less than the number. What is the measure of the first angle?

Let x represent the unknown angle measure. Write an appropriate equation, and solve it for x.

51. Find the complement of a 30° angle.

52. Find the supplement of a 30° angle.

53. Find the supplement of a 105° angle.

54. Find the complement of a 75° angle.

Write an equation to solve each problem.

55. The degree measure of an angle is eight times as large as its complement's measure. What is the measure of the angle?

56. The measure of an angle is 10° more than three times its complement's measure. What is the measure of the angle?

57. The measure of an angle is 10° more than its supplement's measure. What is the measure of the angle?

58. The measure of an angle is 20° less than nineteen times its supplement's measure. What is the measure of the angle?

APPLICATIONS

59. SYNTHESIZER Refer to Illustration 6. Find *x* and *y*.

ILLUSTRATION 6

60. AVIATION Refer to Illustration 7. How many degrees from the horizontal position are the wings of the airplane?

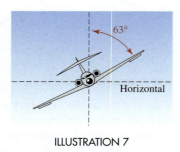

ILLUSTRATION 7

61. GARDENING   In Illustration 8, what angle does the handle of the lawn mower make with the ground?

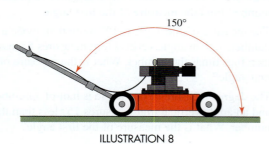

ILLUSTRATION 8

62. ANGLES   Each of the following drawings illustrates a real-life application of a concept that we discussed in this section. Tell what concept is illustrated.

a. Railroad crossing guard

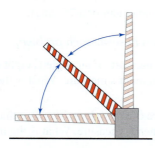

b. Miter saw, used to cut molding

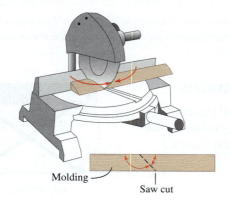

Molding — |
        Saw cut

c. Keyboard stand

WRITING

63. Explain why an angle measuring 105° cannot have a complement.

64. Explain why an angle measuring 210° cannot have a supplement.

65. In parts a and b, explain why the two highlighted angles aren't considered to be adjacent angles.

a.

b.

c. Explain why the angles highlighted below are not vertical angles.

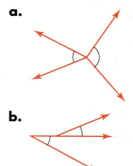

66. Some of the vocabulary introduced in this section can be used in other contexts. In your own words, explain what is meant by each sentence.

a. A homeowner decided to purchase the vacant lot *adjacent* to his property.

b. A teacher decided to *supplement* his income by working a second job on weekends.

c. The right beverage can *complement* a well-cooked meal.

3   *Parallel and Perpendicular Lines*

In this section, you will learn about

- Parallel and perpendicular lines • Transversals and angles
- Properties of parallel lines • Converses

INTRODUCTION. In this section, we will consider *parallel* and *perpendicular* lines. Since parallel lines are always the same distance apart, the railroad tracks shown in Figure 25(a) illustrate one application of parallel lines. Figure 25(b) shows one of the events of men's gymnastics, the parallel bars. Since perpendicular lines meet and form right angles, the monument and the ground shown in Figure 25(c) illustrate one application of perpendicular lines.

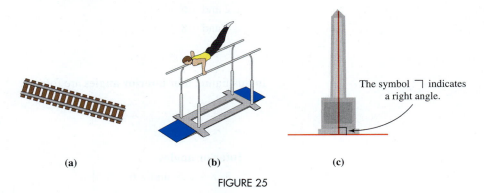

The symbol ⌐ indicates a right angle.

(a)   (b)   (c)

FIGURE 25

Parallel and perpendicular lines

If two lines lie in the same plane, they are called **coplanar.** Two coplanar lines that do not intersect are called **parallel lines.** See Figure 26(a). If two lines do not lie in the same plane, they are called **noncoplanar.** Two noncoplanar lines that do not intersect are called **skew lines.**

Parallel lines

> **Parallel lines** are coplanar lines that do not intersect.

If lines l_1 (read as "l sub 1") and l_2 (read as "l sub 2") are parallel, we can write $l_1 \parallel l_2$, where the symbol $\parallel$ is read as "is parallel to."

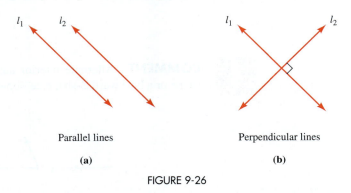

Parallel lines   Perpendicular lines

(a)   (b)

FIGURE 9-26

Perpendicular lines

> **Perpendicular lines** are lines that intersect and form right angles.

In Figure 26(b), $l_1 \perp l_2$, where the symbol $\perp$ is read as "is perpendicular to."

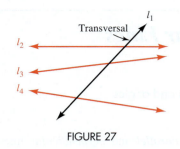

FIGURE 27

Transversals and angles

A line that intersects two or more coplanar lines is called a **transversal.** For example, line l_1 in Figure 27 is a transversal intersecting lines l_2, l_3, and l_4.

When *two* lines are cut by a transversal, all eight of the angles that are formed are important in the study of parallel lines. Descriptive names are given to several pairs of these angles, as shown below.

In this figure, four pairs of **corresponding angles** are formed.

Corresponding angles

$\angle 1$ and $\angle 5$

$\angle 3$ and $\angle 7$

$\angle 2$ and $\angle 6$

$\angle 4$ and $\angle 8$

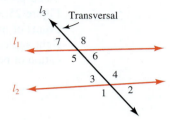

In this figure, four **interior angles** are formed.

Interior angles

$\angle 3$, $\angle 4$, $\angle 5$, and $\angle 6$

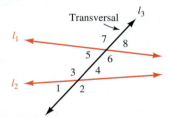

In this figure, two pairs of **alternate interior angles** are formed.

Alternate interior angles

$\angle 4$ and $\angle 5$

$\angle 3$ and $\angle 6$

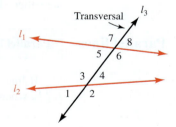

 COMMENT   Alternate interior angles are easily spotted, because they form a Z-shape or a backward Z-shape, as shown in Figure 28.

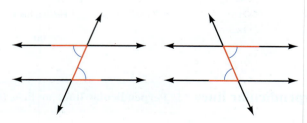

FIGURE 28

EXAMPLE 1   *Identifying angles.*   In Figure 29, identify   **a.** all pairs of alternate interior angles,   **b.** all pairs of corresponding angles,   and   **c.** all interior angles.

Solution   **a.** Pairs of alternate interior angles in the figure are

$\angle 3$ and $\angle 5$, $\angle 4$ and $\angle 6$

b. Pairs of corresponding angles are

$\angle 1$ and $\angle 5$, $\angle 4$ and $\angle 8$,
$\angle 2$ and $\angle 6$, $\angle 3$ and $\angle 7$

c. Interior angles are

$\angle 3$, $\angle 4$, $\angle 5$, and $\angle 6$

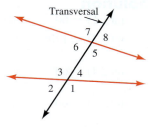

FIGURE 29   ■

Properties of parallel lines

Lines that are cut by a transversal may or may not be parallel. When a pair of parallel lines are cut by a transversal, we can make several important observations about the angles that are formed.

1. If two parallel lines are cut by a transversal, corresponding angles are congruent. (See Figure 30.) If $l_1 \parallel l_2$, then $\angle 1 \cong \angle 5$, $\angle 3 \cong \angle 7$, $\angle 2 \cong \angle 6$, and $\angle 4 \cong \angle 8$.

2. If two parallel lines are cut by a transversal, alternate interior angles are congruent. (See Figure 30.) If $l_1 \parallel l_2$, then $\angle 3 \cong \angle 6$ and $\angle 4 \cong \angle 5$.

3. If two parallel lines are cut by a transversal, interior angles on the same side of the transversal are supplementary. (See Figure 30.) If $l_1 \parallel l_2$, then $\angle 3$ is supplementary to $\angle 5$ and $\angle 4$ is supplementary to $\angle 6$.

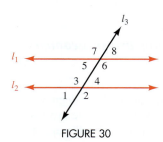

FIGURE 30

4. If a transversal is perpendicular to one of two parallel lines, it is also perpendicular to the other line. (See Figure 31.) If $l_1 \parallel l_2$ and $l_3 \perp l_1$, then $l_3 \perp l_2$.

5. If two lines are parallel to a third line, they are parallel to each other. (See Figure 32.) If $l_1 \parallel l_2$ and $l_1 \parallel l_3$, then $l_2 \parallel l_3$.

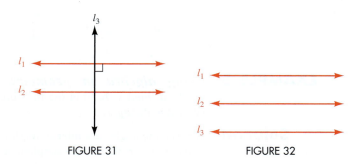

FIGURE 31   FIGURE 32

EXAMPLE 2 *Evaluating angles.* See Figure 33. If $l_1 \parallel l_2$ and m($\angle 3$) = 120°, find the measures of the other angles.

Solution

m($\angle 1$) = 60° $\angle 3$ and $\angle 1$ are supplementary.

m($\angle 2$) = 120°   Vertical angles are congruent: m($\angle 2$) = m($\angle 3$).

m($\angle 4$) = 60°   Vertical angles are congruent: m($\angle 4$) = m($\angle 1$).

m($\angle 5$) = 60°   If two parallel lines are cut by a transversal, alternate interior angles are congruent: m($\angle 5$) = m($\angle 4$).

m($\angle 6$) = 120°   If two parallel lines are cut by a transversal, alternate interior angles are congruent: m($\angle 6$) = m($\angle 3$).

m($\angle 7$) = 120°   Vertical angles are congruent: m($\angle 7$) = m($\angle 6$).

m($\angle 8$) = 60°   Vertical angles are congruent: m($\angle 8$) = m($\angle 5$).

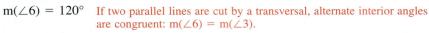

FIGURE 33

Self Check

If $l_1 \parallel l_2$ and m($\angle 8$) = 50°, find the measures of the other angles. (See Figure 33.)

Answers: m($\angle 5$) = 50°, m($\angle 7$) = 130°, m($\angle 6$) = 130°, m($\angle 3$) = 130°, m($\angle 4$) = 50°, m($\angle 1$) = 50°, m($\angle 2$) = 130° ■

EXAMPLE 3 *Two transversals.* See Figure 34. If $\overline{AB} \parallel \overline{DE}$, which pairs of angles are congruent?

Solution Since $\overline{AB} \parallel \overline{DE}$, and $\overleftrightarrow{AC}$ is a transversal cutting them, corresponding angles are congruent. So we have

$$\angle A \cong \angle 1$$

Since $\overline{AB} \parallel \overline{DE}$ and $\overleftrightarrow{BC}$ is a transversal cutting them, corresponding angles must be congruent. So we have

$$\angle B \cong \angle 2$$

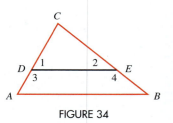

FIGURE 34

■

EXAMPLE 4 *Using algebra in geometry.*
In Figure 35, $l_1 \parallel l_2$. Find x.

Solution
In the figure, two corresponding angles have degree measures that are represented by the algebraic expressions $9x - 15°$ and $6x + 30°$. Since $l_1 \parallel l_2$, all pairs of corresponding angles are congruent.

$9x - 15° = 6x + 30°$ The angle measures are equal.

$3x - 15° = 30°$ To eliminate 6x from the right-hand side, subtract 6x from both sides.

$3x = 45°$ To undo the subtraction of 15°, add 15° to both sides: 30° + 15° = 45°.

$x = 15°$ To undo the multiplication by 3, divide both sides by 3.

Thus, $x = 15°$.

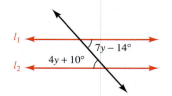

FIGURE 35

Self Check
In the figure below, $l_1 \parallel l_2$. Find y.

Answer: 8° ■

EXAMPLE 5 *Using algebra in geometry.* In Figure 36, $l_1 \parallel l_2$. **a.** Find x. **b.** Find the measures of both angles labeled in the figure.

Solution **a.** Since the angles are interior angles on the same side of the transversal, they are supplementary.

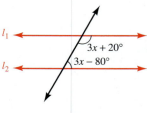

FIGURE 36

$$3x - 80° + 3x + 20° = 180°$$ The sum of the measures of two supplementary angles is 180°.

$$6x - 60° = 180°$$ Combine like terms.

$$6x = 240°$$ To undo the subtraction of 60°, add 60° to both sides: 180° + 60° = 240°.

$$x = 40°$$ To undo the multiplication by 6, divide both sides by 6.

Thus, $x = 40°$.

This problem may be solved using a different approach. In Figure 37, we see that $\angle 1$ and the angle with measure $3x - 80°$ are corresponding angles. Since l_1 and l_2 are parallel, all pairs of corresponding angles are congruent. Therefore,

$$m(\angle 1) = 3x - 80°$$

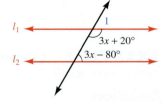

FIGURE 37

In the figure, we also see that $\angle 1$ and the angle with measure $3x + 20°$ are supplementary. That means that the sum of their measures must be 180°. We have

$$m(\angle 1) + 3x + 20° = 180°$$
$$\mathbf{3x - 80°} + 3x + 20° = 180°$$ Replace $m(\angle 1)$ with $3x - 80°$.

This is the same equation that we obtained in the previous solution. When it is solved, we find that $x = 40°$.

b. To find the measures of the angles in Figure 37, we evaluate the expressions $3x + 20°$ and $3x - 80°$ for $x = 40°$.

$$3x + 20° = 3(\mathbf{40°}) + 20° \qquad 3x - 80° = 3(\mathbf{40°}) - 80°$$
$$= 120° + 20° \qquad\qquad\qquad = 120° - 80°$$
$$= 140° \qquad\qquad\qquad\qquad = 40°$$

The measures of the angles labeled in Figure 37 are 140° and 40°.   ■

Converses

Many geometric facts are stated in *if ....., then ....* form. For example, we have seen that

If two angles are vertical angles, then they are congruent.

When studying such statements, it is worthwhile to interchange their parts and then to determine whether the "reverse" is true.

If two angles are congruent, then they are vertical angles.

In this case, the resulting statement is not true.

If a mathematical statement is written in the form *if p..., then q...*, we call the statement *if q..., then p...* its **converse.** It is interesting to note that the converses of some statements are true, while the converses of other statements are false.

Earlier in this section, we saw that

- If two parallel lines are cut by a transversal, then corresponding angles are congruent.

- If two parallel lines are cut by a transversal, then alternate interior angles are congruent.

It can be proved that the converses of these two statements are true. That is, given two lines cut by a transversal,

- If a pair of corresponding angles are congruent, then the lines are parallel.
- If a pair of alternate interior angles are congruent, then the lines are parallel.

When a statement and its converse are both true, we can combine them into a single statement using the phrase *if and only if*.

Properties of parallel lines

Given two lines cut by a transversal,

1. Corresponding angles are congruent if and only if the lines are parallel.
2. Alternate interior angles are congruent if and only if the lines are parallel.

EXAMPLE 6   *Converses.*   In Figure 38, are lines l_1 and l_2 parallel?

Solution

We have two lines that are cut by a transversal. Since a pair of alternate interior angles are congruent (both have measure 50°), the lines are parallel.

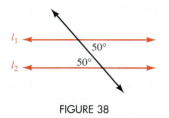

FIGURE 38

Self Check

In the figure below, are lines l_1 and l_2 parallel?

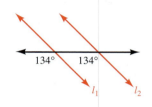

Answer: Since a pair of corresponding angles are congruent, the lines are parallel.

STUDY SET   Section 3

VOCABULARY   *Fill in the blanks.*

1. Two lines that lie in the same plane are _____. Two lines that lie in different planes are _____.

2. Coplanar lines that do not intersect are called _____ lines. Two noncoplanar lines that do not intersect are called _____ lines.

3. If two lines intersect and form right angles, they are _____.

4. A _____ intersects two or more coplanar lines.

5. In Illustration 1, $\angle 4$ and $\angle 6$ are _____ interior angles. $\angle 2$ and $\angle 6$ are _____ angles.

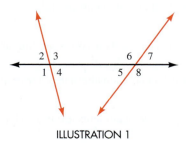

ILLUSTRATION 1

6. If a mathematical statement is written in the form *if p...*, *then q...*, we call the statement *if q..., then p...* its _____.

CONCEPTS

7. **a.** Draw two parallel lines.

   b. Draw two lines that are not parallel.

8. **a.** Draw two perpendicular lines.

b. Draw two lines that are not perpendicular.

9. a. Draw two parallel lines cut by a transversal.

b. Draw two lines that are not parallel cut by a transversal.

10. Draw three parallel lines.

In Exercises 11–13, two parallel lines are cut by a transversal. Fill in the blanks.

11. In Illustration 2, we know that $\angle 1$ _____ $\angle 2$, because when two parallel lines are cut by a transversal, _____ _____ angles are congruent.

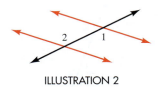

ILLUSTRATION 2

12. In Illustration 3, we know that $\angle ABC$ _____ $\angle BEF$, because when two parallel lines are cut by a transversal, _____ angles are congruent.

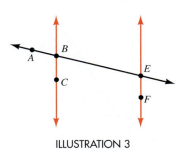

ILLUSTRATION 3

13. In Illustration 4, we know that m($\angle ABC$) + m($\angle BCD$) = _____, because when two parallel lines are cut by a transversal, _____ angles on the same side of the transversal are supplementary.

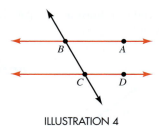

ILLUSTRATION 4

14. a. Explain why l_1 and l_2 in Illustration 5 are parallel.

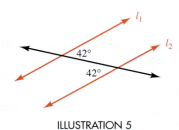

ILLUSTRATION 5

b. Explain why l_1 and l_2 in Illustration 6 are parallel.

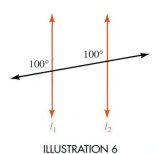

ILLUSTRATION 6

c. Explain why l_1 and l_2 in Illustration 7 are not parallel.

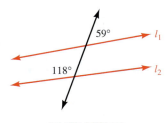

ILLUSTRATION 7

15. Are lines l_1 and l_2 in Illustration 8 parallel? Explain your answer.

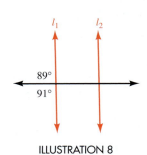

ILLUSTRATION 8

16. Are lines l_1 and l_2 in Illustration 9 parallel? Explain why or why not.

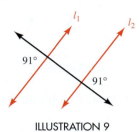

ILLUSTRATION 9

17. a. Which pairs of angles in Illustration 10 are alternate interior angles?
   b. Which pairs of angles in Illustration 10 are corresponding angles?
   c. Which angles in Illustration 10 are interior angles?

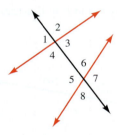

ILLUSTRATION 10

18. In Illustration 11, $l_1 \parallel l_2$. What can you conclude about l_1 and l_3?

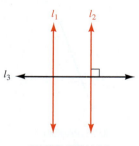

ILLUSTRATION 11

19. In Illustration 12, $l_1 \parallel l_2$ and $l_2 \parallel l_3$. What can you conclude about l_1 and l_3?

ILLUSTRATION 12

20. In Illustration 13, $\overline{AB} \parallel \overline{DE}$. What pairs of angles are congruent? Explain your reasoning.

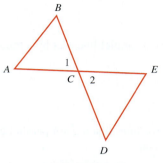

ILLUSTRATION 13

NOTATION   *Fill in the blanks.*

21. The symbol ⌐ indicates _____.
22. The symbol $\parallel$ is read as "_____."
23. The symbol $\perp$ is read as "_____."
24. The symbol l_1 is read as "_____."

PRACTICE

25. In Illustration 14, $l_1 \parallel l_2$ and $m(\angle 4) = 130°$. Find the measures of the other angles.

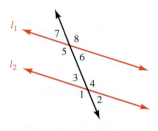

ILLUSTRATION 14

26. In Illustration 15, $l_1 \parallel l_2$ and $m(\angle 2) = 40°$. Find the measures of the other angles.

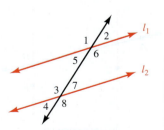

ILLUSTRATION 15

27. In Illustration 16, on the next page, $l_1 \parallel \overline{AB}$.
   a. Find $m(\angle 1)$, $m(\angle 2)$, $m(\angle 3)$, and $m(\angle 4)$.
   b. Find $m(\angle 1) + m(\angle 2) + m(\angle ACD)$.
   c. Find $m(\angle 1) + m(\angle ABC) + m(\angle 4)$.

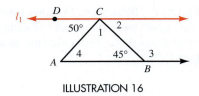

ILLUSTRATION 16

28. In Illustration 17, $\overline{AB} \parallel \overline{DE}$. Find m($\angle B$), m($\angle E$), and m($\angle 1$).

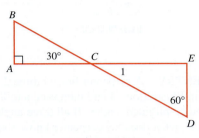

ILLUSTRATION 17

In Exercises 29–32, $l_1 \parallel l_2$. First find x. Then determine the measure of each angle that is labeled in the figure.

29.

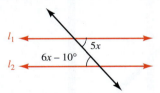

30.

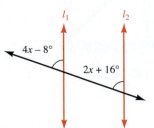

31.

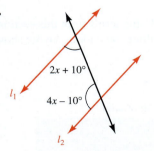

32.

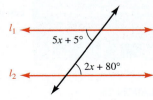

In Exercises 33–36, first find x. Then determine the measure of each angle that is labeled in the figure.

33. $l_1 \parallel \overline{CA}$

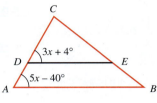

34. $\overline{AB} \parallel \overline{DE}$

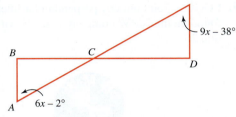

35. $\overline{AB} \parallel \overline{DE}$

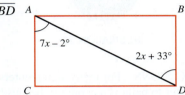

36. $\overline{AC} \parallel \overline{BD}$

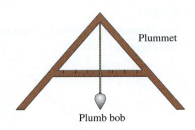

APPLICATIONS

37. CONSTRUCTING PYRAMIDS   The Egyptians used a device called a **plummet** to tell whether stones were properly leveled. A plummet (shown in Illustration 18) is made up of an A-frame and a plumb bob suspended from the peak of the frame. How could a builder use a plummet to tell that the stone on the left is not level and that the stones on the right are level?

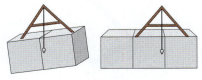

ILLUSTRATION 18

38. DIAGRAMMING SENTENCES   English instructors have their students diagram sentences to help teach proper sentence structure. Illustration 19 is a diagram of the sentence *The cave was rather dark and damp.* Point out pairs of parallel and perpendicular lines used in the diagram.

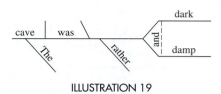

ILLUSTRATION 19

39. LOGO   Point out any perpendicular lines that can be found on the BMW company logo shown in Illustration 20.

ILLUSTRATION 20

40. PAINTING SIGNS   For many sign painters, the most difficult letter to paint is a capital E, because of all of the right angles involved. See Illustration 21. How many right angles are there?

ILLUSTRATION 21

41. HANGING WALLPAPER   Explain why the concepts of perpendicular and parallel are both important when hanging wallpaper.

42. TOOLS   See Illustration 22. What geometric concepts are seen in the design of the rake?

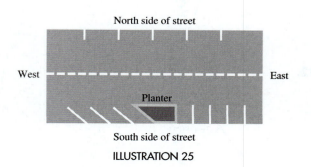

ILLUSTRATION 22

43. SEISMOLOGY   Illustration 23 shows how an earthquake fault occurs when two blocks of earth move apart and one part drops down. Determine the measures of ∠1, ∠2, and ∠3.

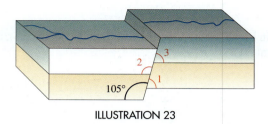

ILLUSTRATION 23

44. CARPENTRY   A carpenter braced three 2 × 4's as shown in Illustration 24 and then used a tool to measure the three highlighted angles. If all three angles measured 45°, what does the carpenter know about the three 2 × 4's? Explain your answer.

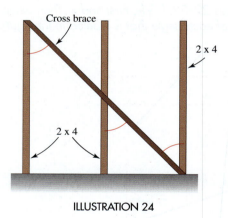

ILLUSTRATION 24

WRITING

45. PARKING DESIGN   Using terms from this section, write a paragraph describing the parking layout shown in Illustration 25.

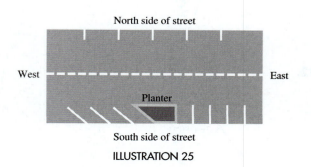

ILLUSTRATION 25

46. In your own words, explain what is meant by each of the following sentences.
  a. The hikers were told that the path *parallels* the river.
  b. John's quick rise to fame and fortune *paralleled* that of his older brother.
  c. The judge stated that the case that was before her court was without *parallel*.

47. Are pairs of alternate interior angles always congruent? Explain.

48. Are pairs of interior angles always supplementary? Explain.

---

4 Triangles

In this section, you will learn about

- Polygons • Triangles • Classifying triangles • Properties of isosceles triangles
- The sum of the measures of the angles of a triangle

INTRODUCTION. We will now discuss geometric figures called *polygons*. We see these shapes every day. For example, the walls of most buildings are rectangular in shape. Some tile and vinyl floor patterns use the shape of a pentagon or a hexagon. Stop signs are in the shape of an octagon. In this section, we will focus on one specific type of polygon called a *triangle*. Triangular shapes are especially important because triangles contribute strength and stability to walls and towers. The gable ends of many houses are triangular, as are the sides of the Great Pyramid of Egypt.

Polygons

Polygon

> A **polygon** is a closed geometric figure with at least three line segments for its sides.

Polygons are formed by fitting together line segments in such a way that

- the segments only intersect at their endpoints,
- no two line segments with a common endpoint lie on the same line, and
- each endpoint is shared by exactly two segments.

The line segments that form a polygon are called its **sides.** The point where two sides of a polygon intersect is called a **vertex** of the polygon (plural **vertices**). The polygon shown in Figure 39 has 5 sides and 5 vertices.

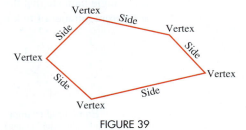

FIGURE 39

Polygons are classified according to the number of sides that they have. For example, in Figure 40 (on the next page), we see that a polygon with four sides is called a *quadrilateral,* and a polygon with eight sides is called an *octagon*. If a polygon has sides that are all the same length and angles that are the same measure, we call it a **regular polygon.**

| Triangle
3 sides | Quadrilateral
4 sides | Pentagon
5 sides | Hexagon
6 sides | Heptagon
7 sides | Octagon
8 sides | Nonagon
9 sides | Decagon
10 sides | Dodecagon
12 sides |

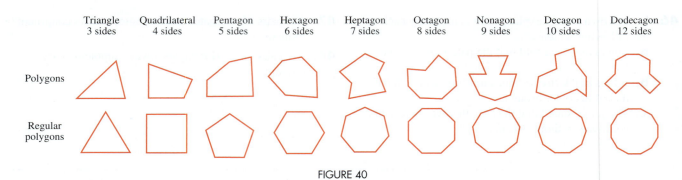

FIGURE 40

EXAMPLE 1 *Vertices of a polygon.* Give the number of vertices of
a. a triangle   and   **b.** a hexagon.

Solution

a. From Figure 40, we see that a triangle has three angles and therefore three vertices.

b. From Figure 40, we see that a hexagon has six angles and therefore six vertices.

Self Check

Give the number of vertices of

a. a quadrilateral

b. a pentagon

Answer: **a.** 4 **b.** 5 ∎

From the results of Example 1, we see that the number of vertices of a polygon is equal to the number of its sides. The number of its vertices is also equal to the number of its angles.

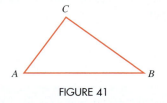

FIGURE 41

Triangles

A **triangle** is a polygon with three sides (and three vertices). Recall that in geometry points are labeled with capital letters. We can use the capital letters that denote the vertices of a triangle to name the triangle. For example, when referring to the triangle in Figure 41, with vertices *A, B,* and *C,* we can use the notation △*ABC* (read as "triangle *ABC*").

COMMENT   When naming a triangle, we may begin with any vertex. Then we move around the figure in a clockwise (or counterclockwise) direction as we list the remaining vertices. Other ways of naming the triangle in Figure 41 are △*ACB*, △*BCA*, △*BAC*, △*CAB*, and △*CBA*.

Classifying triangles

Figure 42 shows how triangles can be classified according to the lengths of their sides. The single tick marks drawn on each side of the equilateral triangle indicate that the sides are of equal length. The double tick marks drawn on two of the sides of the isosceles triangle indicate that they have the same length. Each side of the scalene triangle has a different number of tick marks to indicate that the sides have different lengths.

Equilateral triangle
(all sides equal length)

Isosceles triangle
(at least two sides of
equal length)

Scalene triangle
(no sides equal length)

FIGURE 42

COMMENT   Since equilateral triangles have at least two sides of equal length, they are also isosceles. However, isosceles triangles are not necessarily equilateral.

Triangles may also be classified by their angles, as shown in Figure 43.

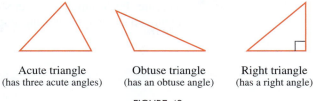

Acute triangle   Obtuse triangle   Right triangle
(has three acute angles)   (has an obtuse angle)   (has a right angle)

FIGURE 43

Right triangles have many real-life applications. For example, in Figure 44(a), we see that a right triangle is formed when a ladder leans against the wall of a building. The longest side of a right triangle is called the **hypotenuse,** and the other two sides are called **legs.** The hypotenuse of a right triangle is always opposite the 90° (right) angle. The legs of a right triangle are adjacent to (next to) the right angle. See Figure 44(b).

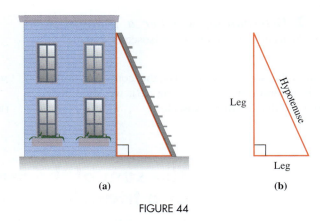

(a)   (b)

FIGURE 44

Properties of isosceles triangles

In an isosceles triangle, the angles opposite the sides of equal length are called **base angles,** the sides of equal length form the **vertex angle,** and the third side is called the **base.** Two isosceles triangles are shown in Figure 45.

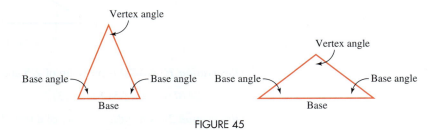

FIGURE 45

We have seen that isosceles triangles have two sides of equal length. The **isosceles triangle theorem** states that such triangles have one other important characteristic: Their base angles are congruent.

Isosceles triangle theorem

If two sides of a triangle are congruent, then the angles opposite those sides are congruent.

COMMENT   Tick marks can be used to denote the sides of a triangle that have the same length. They can also be used to indicate the angles of a triangle with the same measure. For example, we can show that the base angles of the isosceles triangle in Figure 46 are congruent using single tick marks.

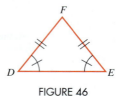

$\angle D$ is opposite $\overline{FE}$, and $\angle E$ is opposite $\overline{FD}$. By the isosceles triangle theorem, if m$(\overline{FD})$ = m$(\overline{FE})$, then m$(\angle D)$ = m$(\angle E)$.

FIGURE 46

The *converse* of the isosceles triangle theorem is also true.

Converse of the isosceles triangle theorem	If two angles of a triangle are congruent, then the sides opposite the angles have the same length, and the triangle is isosceles.

EXAMPLE 2 *Determining whether a triangle is isosceles.*   Is the triangle in Figure 47 an isosceles triangle?

Solution

$\angle A$ and $\angle B$ have the same measure. By the converse of the isosceles triangle theorem, if m$(\angle A)$ = m$(\angle B)$, we know that m$(\overline{BC})$ = m$(\overline{AC})$ and that $\triangle ABC$ is isosceles.

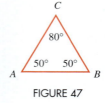

FIGURE 47

Self Check

Is the triangle shown below an isosceles triangle?

Answer: no

The sum of the measures of the angles of a triangle

To prove a very important fact about triangles, we begin with $\triangle ABC$ as shown in Figure 48. Line *l* is drawn through point *B* so that it is parallel to $\overline{AC}$. For easy reference, several of the angles are labeled with numbers. Three observations from the figure will be helpful in the proof.

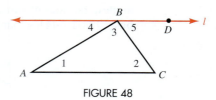

FIGURE 48

Observation 1: $\angle ABD$ is made up of $\angle 5$ and $\angle 3$. Therefore,

$$\text{m}(\angle ABD) = \text{m}(\angle 5) + \text{m}(\angle 3)$$

Observation 2:   Since the measure of a straight angle is 180°,

$$\text{m}(\angle ABD) + \text{m}(\angle 4) = 180°$$

Observation 3:   Since $\overleftrightarrow{AB}$ is a transversal cutting parallel lines $\overleftrightarrow{AC}$ and *l*, alternate interior angles are congruent, and therefore

$$\text{m}(\angle 1) = \text{m}(\angle 4)$$

Similarly, since $\overleftrightarrow{BC}$ is a transversal cutting parallel lines $\overleftrightarrow{AC}$ and *l*, alternate interior angles are congruent, and therefore

$$\text{m}(\angle 2) = \text{m}(\angle 5)$$

Proof   To begin the proof, we apply the addition property of equality to the equation from Observation 1.

$$m(\angle ABD) = m(\angle 5) + m(\angle 3)$$

$$m(\angle ABD) + \mathbf{m(\angle 4)} = \mathbf{m(\angle 4)} + m(\angle 5) + m(\angle 3) \quad \text{Add } m(\angle 4) \text{ to both sides.}$$

From Observation 2, we can replace $m(\angle ABD) + m(\angle 4)$ with 180°.

$$180° = \mathbf{m(\angle 4)} + \mathbf{m(\angle 5)} + m(\angle 3)$$

From Observation 3, we can replace $m(\angle 4)$ with $m(\angle 1)$, and $m(\angle 5)$ with $m(\angle 2)$.

$$180° = \mathbf{m(\angle 1)} + \mathbf{m(\angle 2)} + m(\angle 3)$$

The result indicates that the sum of the angles of $\triangle ABC$ in Figure 48 is 180°. Because of the generality of this proof, we have shown this fact to be true for any triangle. ☐

Angles of a triangle

> The sum of the angle measures of any triangle is 180°.

If you draw several triangles and carefully measure each angle with a protractor, you will find that the sum of the angle measures of each triangle is 180°. See Figure 49.

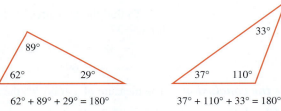

$$62° + 89° + 29° = 180° \qquad 37° + 110° + 33° = 180°$$

FIGURE 49

EXAMPLE 3   *Sum of the angles of a triangle.*   See Figure 50. Find x.

Solution

Since the sum of the angle measures of any triangle is 180°, we have

$x + 40° + 90° = 180°$ The ⌐ symbol indicates that an angle has a measure of 90°.

$x + 130° = 180°$ Do the addition: $40° + 90° = 130°$.

$x = 50°$ To undo the addition of 130°, subtract 130° from both sides.

Self Check

In the figure below, find y.

Answer: 90°   ∎

FIGURE 50

EXAMPLE 4   *Sum of the angles of a triangle.*   In $\triangle ABC$, the measure of $\angle A$ exceeds the measure of $\angle B$ by 32°, and the measure $\angle C$ is twice the measure of $\angle B$. Find the measure of each angle of $\triangle ABC$.

Analyze the problem
- There are three angles to consider: $\angle A$, $\angle B$, and $\angle C$.
- The measure of $\angle A$ exceeds the measure of $\angle B$ by 32°.
- The measure of $\angle C$ is twice the measure of $\angle B$.
- We are to find the measure of each angle.

Form an equation   Since the measures of $\angle A$ and $\angle C$ are related to the measure of $\angle B$, we begin by letting $x =$ the measure of $\angle B$. Then we translate the words *exceeds by 32°* and *twice*, to get algebraic expressions that represent the measures of the other two angles.

$$x + 32° = \text{the measure of } \angle A$$
$$2x = \text{the measure of } \angle C$$

At this stage, drawing a sketch of the situation is often helpful. See Figure 51.

Because the sum of the measures of the angles of any triangle is 180°,

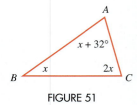

FIGURE 51

The measure of $\angle A$	plus	the measure of $\angle B$	plus	the measure of $\angle C$	is	180°.
$x + 32°$	$+$	x	$+$	$2x$	$=$	$180°$

Solve the equation

$$x + 32° + x + 2x = 180°$$

$$4x + 32° = 180° \qquad \text{Combine like terms: } x + x + 2x = 4x.$$

$$4x + 32° - \mathbf{32°} = 180° - \mathbf{32°} \qquad \text{To undo the addition of 32°, subtract 32° from both sides.}$$

$$4x = 148° \qquad \text{Do the subtractions.}$$

$$\frac{4x}{\mathbf{4}} = \frac{148°}{\mathbf{4}} \qquad \text{To undo the multiplication by 4, divide both sides by 4.}$$

$$x = 37° \qquad \text{Do the divisions.}$$

To find the measures of $\angle A$ and $\angle C$, we evaluate the expressions $x + 32°$ and $2x$ for $x = 37°$.

$$x + 32° = \mathbf{37°} + 32° \quad \text{Substitute 37 for } x. \qquad 2x = 2(\mathbf{37°}) \quad \text{Substitute 37 for } x.$$
$$= 69° \qquad\qquad\qquad\qquad\qquad = 74°$$

State the conclusion   The measure of $\angle B$ is 37°, the measure of $\angle A$ is 69°, and the measure of $\angle C$ is 74°.

Check the result   First, we note that 69° exceeds 37° by 32° and that 74° is twice 37°. Furthermore, since $37° + 69° + 74° = 180°$, the results check.   ■

EXAMPLE 5   ***Vertex angle of an isosceles triangle.***   See Figure 52. If one base angle of an isosceles triangle measures 70°, how large is the vertex angle?

Solution

By the isosceles triangle theorem, if one of the base angles measures 70°, so does the other. If we let x represent the measure of the vertex angle, we have

FIGURE 52

$$x + 70° + 70° = 180° \quad \text{The sum of the measures of the angles of a triangle is 180°.}$$
$$x + 140° = 180° \quad \text{Combine like terms: } 70° + 70° = 140°.$$
$$x = 40° \quad \text{To undo the addition of 140°, subtract 140° from both sides.}$$

The vertex angle measures 40°.   ■

EXAMPLE 6   ***Base angles of an isosceles triangle.***   See Figure 53. If the vertex angle of an isosceles triangle measures 99°, what are the measures of the base angles?

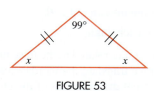

FIGURE 53

Self Check

Find the measures of the base angles of the isosceles triangle shown below.

Solution

The base angles of an isosceles triangle have the same measure. If we let x represent the measure of one base angle, the measure of the other base angle is also x. See Figure 53. Since the sum of the measures of the angles of any triangle is 180°, the sum of the measures of the base angles and of the vertex angle is 180°. We can use this fact to form an equation.

$x + x + 99° = 180°$

$2x + 99° = 180°$ Combine like terms: $x + x = 2x$.

$2x = 81°$ To undo the addition of 99°, subtract 99° from both sides.

$\dfrac{2x}{2} = \dfrac{81°}{2}$ To undo the multiplication by 2, divide both sides by 2.

$x = 40.5°$

The measure of each base angle is 40.5°.

Answer: 61.5° ∎

STUDY SET   Section 4

VOCABULARY   *Fill in the blanks.*

1. A _____ is a closed geometric figure with at least three line segments for its sides.

2. The polygon shown in Illustration 1 has seven _____ and seven _____.

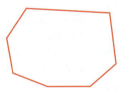

ILLUSTRATION 1

3. A point where two sides of a polygon intersect is called a _____ of the polygon.

4. A _____ polygon has sides that are all the same length and angles that all have the same measure.

5. A triangle with three sides of equal length is called an _____ triangle. An _____ triangle has at least two sides of equal length. A _____ triangle has no sides of equal length.

6. An _____ triangle has three acute angles. An _____ triangle has one obtuse angle. A _____ triangle has one right angle.

7. The longest side of a right triangle is the _____. The other two sides of a right triangle are called _____.

8. The _____ angles of an isosceles triangle have the same measure. The sides of equal length of an isosceles triangle form the _____ angle.

9. In this section, we discussed the sum of the measures of the angles of a triangle. The word *sum* indicates the operation of _____.

10. Complete the table.

Number of sides	Name of polygon
3	
4	
5	
6	
7	
8	
9	
10	
12	

CONCEPTS

11. Draw an example of each type of polygon.
   a. hexagon   **b.** octagon
   c. quadrilateral **d.** triangle
   e. pentagon   **f.** decagon

12. For each polygon, give the number of sides it has, tell its name, and then give the number of vertices it has.

a.   **b.**

c.   **d.**

e.   **f.**

g.

h.

i.

j.

13. Explain why the given polygon is not a regular polygon.

a.

b.

14. Refer to the triangle in Illustration 2.
   a. What are the names of the vertices of the triangle?
   b. Use the vertices to name this triangle in three ways.
   c. How many sides does the triangle have? Name them.

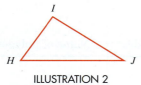

ILLUSTRATION 2

15. Draw an example of each type of triangle.
   a. Isosceles   **b.** Equilateral   **c.** Scalene

16. Classify each triangle as an equilateral triangle, an isosceles triangle, or a scalene triangle.
   a.

   b.

   c.

   d.

   e.

   f.

   g.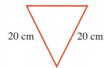

   h.

17. Draw an example of each type of triangle.
   a. Obtuse   **b.** Right   **c.** Acute

18. Classify each triangle as an acute, an obtuse, or a right triangle.

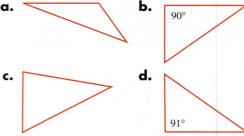

19. Refer to the triangle shown in Illustration 3.
   a. What type of triangle is it?
   b. What is the measure of ∠B?
   c. What two line segments form the legs?
   d. What line segment is the hypotenuse?
   e. Which side of the triangle is the longest?
   f. Which side is opposite ∠B?

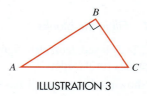

ILLUSTRATION 3

20. Fill in the blanks.
   a. The sides of a right triangle that are adjacent to the right angle are called the _____.
   b. The hypotenuse of a right triangle is the side _____ the right angle.

21. Fill in the blanks.
   a. The _____ triangle theorem states that if two sides of a triangle are congruent, then the _____ opposite those sides are congruent.
   b. The _____ of the isosceles triangle theorem states that if two angles of a triangle are congruent, then the sides opposite the angles have the same _____, and the triangle is isosceles.

22. Refer to Illustration 4.
   a. What type of triangle is △XYZ?
   b. What two sides are of equal length?
   c. Name the base angles.
   d. Which side is opposite ∠X?
   e. What is the vertex angle?
   f. Which angle is opposite side $\overline{XY}$?
   g. Which two angles are congruent?

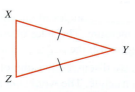

ILLUSTRATION 4

23. Refer to Illustration 5.
 a. What do we know about $\overline{EF}$ and $\overline{GF}$?
 b. What type of triangle is $\triangle EFG$?

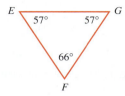

ILLUSTRATION 5

24. a. Find the sum of the measures of the angles of $\triangle JKL$, shown in Illustration 6(a).
 b. Find the sum of the measures of the angles of $\triangle CDE$, shown in Illustration 6(b).

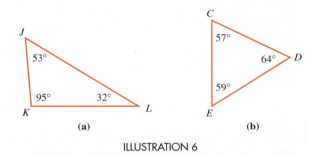

(a)   (b)

ILLUSTRATION 6

25. What is the sum of the measures of the angles of any triangle?
26. Suppose we let x = the measure of $\angle B$.
 a. If the measure of $\angle C$ is twice the measure of $\angle B$, how can we represent the measure of $\angle C$?
 b. If the measure of $\angle D$ is 5° less than four times the measure of $\angle B$, how can we represent the measure of $\angle D$?

NOTATION *Fill in the blanks.*

27. The symbol $\triangle$ means _____.
28. The symbol $m(\angle A)$ means the _____ of angle A.
29. The symbol $m(\overline{AB})$ means the measure of line _____ AB.
30. The symbol $\ulcorner$ means _____ angle.
31. Refer to Illustration 7. What fact about the sides of $\triangle ABC$ do the tick marks indicate?
32. Refer to Illustration 7. What fact about the angles of $\triangle ABC$ do the tick marks indicate?

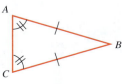

ILLUSTRATION 7

PRACTICE *The measures of two angles of $\triangle ABC$ (shown in Illustration 8) are given. Find the measure of the third angle.*

33. $m(\angle A) = 30°$ and $m(\angle B) = 60°$
 $m(\angle C) = $ _____
34. $m(\angle A) = 45°$ and $m(\angle C) = 105°$
 $m(\angle B) = $ _____
35. $m(\angle B) = 100°$ and $m(\angle A) = 35°$
 $m(\angle C) = $ _____
36. $m(\angle B) = 33°$ and $m(\angle C) = 77°$
 $m(\angle A) = $ _____
37. $m(\angle A) = 25.5°$ and $m(\angle B) = 63.8°$
 $m(\angle C) = $ _____
38. $m(\angle B) = 67.25°$ and $m(\angle C) = 72.5°$
 $m(\angle A) = $ _____

ILLUSTRATION 8

In Exercises 39–40, find y.

39.   **40.**

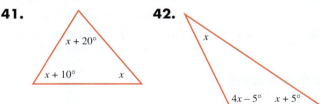

In Exercises 41–44, the degree measures of the angles of a triangle are represented by algebraic expressions. First, find x. Then determine the measure of each angle of the triangle.

41.   **42.**

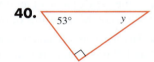

43.   **44.**

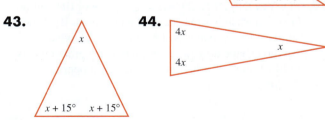

45. The measure of one angle of a triangle is twice that of another angle. The third angle measures 60°. Find the measure of each angle of the triangle.
46. One angle of a triangle has a measure that is 10° more than twice that of a second angle. The third angle has measure 10° less than the first angle. Find the measure of each angle of the triangle.

In Exercises 47–48, find the measure of the vertex angle.

47.

48.

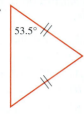

49. The measure of a base angle of an isosceles triangle is 56°. What is the measure of the vertex angle?

50. The measure of a base angle of an isosceles triangle is 85.5°. What is the measure of the vertex angle?

In Exercises 51–52, find x.

51.

52.

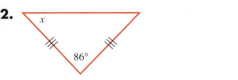

53. If the vertex angle of an isosceles triangle measures 102°, what are the measures of the base angles?

54. If the vertex angle of an isosceles triangle measures 3°, what are the measures of the base angles?

55. The measure of each base angle of an isosceles triangle is 10° less than twice the measure of the vertex angle. Find the measure of each angle of the triangle.

56. The measure of a base angle of an isosceles triangle is 5° more than eight times the measure of the vertex angle. Find the measure of each angle of the triangle.

57. One angle of an isosceles triangle has a measure of 39°. What are the possible measures of the other angles?

58. One angle of an isosceles triangle has a measure of 2°. What are the possible measures of the other angles?

APPLICATIONS

59. POLYGONS IN NATURE   As we see in Illustration 9(a), a starfish has the shape of a pentagon. By drawing similar line segments as shown in part (a), what polygon shape do you see in each of the other objects in Illustration 9?   **b.** Lemon   **c.** Chili pepper   **d.** Apple

(a)   **(b)**

ILLUSTRATION 9

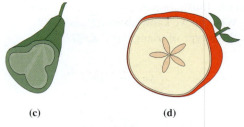

(c)   **(d)**

ILLUSTRATION 9 (continued)

60. CHEMISTRY   Polygons are used to represent the chemical structure of compounds graphically. In Illustration 10, what types of polygons are used to represent methylprednisolone, the active ingredient in an anti-inflammatory medication?

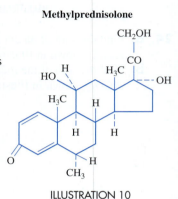

Methylprednisolone

ILLUSTRATION 10

61. AUTOMOBILE JACK   Refer to Illustration 11. Show that no matter how high the jack is raised, it always forms two isosceles triangles.

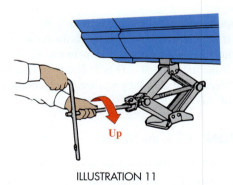

ILLUSTRATION 11

62. EASEL   In Illustration 12, show how two of the legs of the easel form the equal sides of an isosceles triangle.

ILLUSTRATION 12

63. POOL   The rack shown in Illustration 13 is used to set up the billiard balls when beginning a game of pool. Although it does not meet the strict definition of a polygon, the rack has a shape much like a type of triangle discussed in this section. Which type of triangle?

ILLUSTRATION 13

64. DRAFTING   Among the tools used in drafting are the two clear plastic triangles shown in Illustration 14. Classify each according to the lengths of their sides and then according to their angle measures.

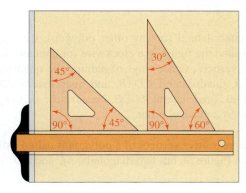

ILLUSTRATION 14

WRITING

65. In this section, we discussed the definition of a pentagon. What is *the* Pentagon? Why is it named that?

66. A student cut a triangular shape out of blue construction paper and labeled the angles ∠1, ∠2, and ∠3, as shown in Illustration 15(a). Then she tore off each of the three corners and arranged them as shown in Illustration 15(b). Explain what important geometric concept this model illustrates.

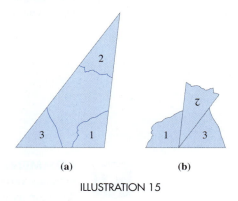

(a)   (b)

ILLUSTRATION 15

67. Explain why a triangle cannot have two right angles.
68. Explain why a triangle cannot have two obtuse angles.

5 *Quadrilaterals and Other Polygons*

In this section, you will learn about

• Quadrilaterals • Properties of rectangles • Trapezoids • The sum of the measures of the angles of a polygon • Finding an angle measure of a regular polygon • Exterior angles of a regular polygon

INTRODUCTION. Recall that a polygon is a closed geometric figure with at least three line segments for its sides. In this section, we will focus on polygons with four sides, called *quadrilaterals*. One type of quadrilateral is the *square*. The game boards for Monopoly and Scrabble have a square shape. Another type of quadrilateral is the *rectangle*. Most picture frames and many mirrors are rectangular. Utility knife blades and swimming fins have shapes that are examples of a third type of quadrilateral called a *trapezoid*.

Quadrilaterals

A **quadrilateral** is a polygon with four sides. Some common quadrilaterals are shown in Figure 54.

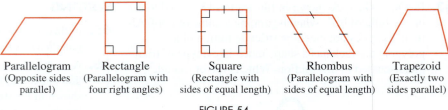

Parallelogram	Rectangle	Square	Rhombus	Trapezoid
(Opposite sides parallel)	(Parallelogram with four right angles)	(Rectangle with sides of equal length)	(Parallelogram with sides of equal length)	(Exactly two sides parallel)

FIGURE 54

We can use the capital letters that denote the vertices of a quadrilateral to name it. For example, when referring to the quadrilateral in Figure 55, with vertices *A, B, C,* and *D,* we can use the notation quadrilateral *ABCD.*

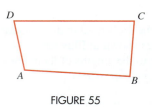

FIGURE 55

COMMENT   When naming a quadrilateral (or any other polygon), we may begin with any vertex. Then we move around the figure in a clockwise (or counterclockwise) direction as we list the remaining vertices. Other ways of naming the quadrilateral in Figure 55 are quadrilateral *ADCB,* quadrilateral *CDAB,* and quadrilateral *DABC.* It would be unacceptable to name it as quadrilateral *ACDB,* because the vertices would not be listed in clockwise (or counterclockwise) order.

A segment that joins two nonconsecutive vertices of a polygon is called a **diagonal** of the polygon. The quadrilateral in Figure 56 has two diagonals, $\overline{AC}$ and $\overline{BD}$.

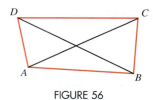

FIGURE 56

Properties of rectangles

Recall that a **rectangle** is a quadrilateral with four right angles. The rectangle is probably the most common and recognizable of all geometric figures. For example, most doors and windows are rectangular in shape. The boundaries of football fields, soccer fields, and basketball courts are rectangles. Even our paper currency, such as the $1, $5, and $20 bills, is in the shape of a rectangle. Rectangles have several important characteristics.

Properties of rectangles

In any rectangle:

1. All four angles are right angles.
2. Opposite sides are parallel.
3. Opposite sides have equal length.
4. The diagonals have equal length.
5. The diagonals intersect at their midpoints.

EXAMPLE 1 *Properties of rectangles.* In Figure 57, quadrilateral *WXYZ* is a rectangle. Find each measure: **a.** m(∠*YXW*), **b.** m($\overline{XY}$), **c.** m($\overline{WY}$), and **d.** m($\overline{XZ}$).

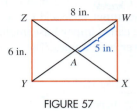

FIGURE 57

Solution

a. In any rectangle, all four angles are right angles. Therefore, ∠*YXW* is a right angle, and m(∠*YXW*) = 90°.

b. $\overline{XY}$ and $\overline{WZ}$ are opposite sides of the rectangle, so they have equal length. Since the length of $\overline{WZ}$ is 8 inches, m($\overline{XY}$) is also 8 inches.

c. $\overline{WY}$ and $\overline{ZX}$ are diagonals of the rectangle, and they intersect at their midpoints. That means that point A is the midpoint of $\overline{WY}$. Since the length of $\overline{WA}$ is 5 inches, m($\overline{WY}$) is 2 · 5 inches or 10 inches.

d. The diagonals of a rectangle are of equal length. In part c, we found that the length of $\overline{WY}$ is 10 inches. Therefore, m($\overline{XZ}$) is also 10 inches.

Self Check

In rectangle *RSTU* shown below, the length of $\overline{RT}$ is 13 ft. Find each measure: **a.** m(∠*SRU*), **b.** m($\overline{ST}$), **c.** m($\overline{TG}$), and **d.** m($\overline{SG}$).

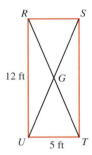

Answers: **a.** 90°, **b.** 12 ft, **c.** 6.5 ft, **d.** 6.5 ft

We have seen that if a quadrilateral has four right angles, it is a rectangle. The following theorems establish some conditions that a parallelogram must meet to ensure that it is a rectangle.

Parallelogram theorems

1. If a parallelogram has one right angle, then the parallelogram is a rectangle.
2. If the diagonals of a parallelogram are congruent, then the parallelogram is a rectangle.

EXAMPLE 2 *Right angle corners.* A carpenter intends to build a shed with a 9-foot-by-12-foot base. How can he make sure that the foundation has four right-angle corners?

Solution

The four-sided foundation, which we will label as quadrilateral *ABCD,* has opposite sides of equal length. See Figure 58. The carpenter can use a tape measure to find the lengths of the diagonals $\overline{AC}$ and $\overline{BD}$. If these diagonals are of equal length, the foundation will be a rectangle and have right angles at its four corners. This process is commonly referred to as "squaring a foundation." Picture framers use a similar process to make sure their frames have four 90° corners.

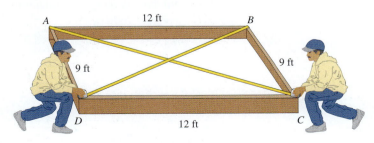

FIGURE 58

Trapezoids

A **trapezoid** is a quadrilateral with exactly two sides parallel. See Figure 59. The parallel sides, in this case $\overline{AB}$ and $\overline{DC}$, are called **bases.** To distinguish between the two bases, we will refer to $\overline{AB}$ as the **upper base** and $\overline{DC}$ as the **lower base.** The angles on either side of the upper base are called **upper base angles,** and the angles on either side of the lower base are called **lower base angles.** The nonparallel sides are called **legs.**

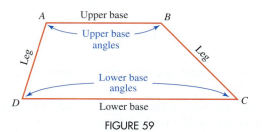

FIGURE 59

In Figure 59, we can view $\overleftrightarrow{AD}$ as a transversal cutting the parallel lines $\overleftrightarrow{AB}$ and $\overleftrightarrow{DC}$. Since $\angle A$ and $\angle D$ are interior angles, they are supplementary. Similarly, $\overleftrightarrow{BC}$ is a transversal cutting the parallel lines $\overleftrightarrow{AB}$ and $\overleftrightarrow{DC}$. Since $\angle B$ and $\angle C$ are interior angles, they are also supplementary. These observations lead us to the conclusion that *there are always two pairs of supplementary angles in any trapezoid.*

EXAMPLE 3   *Angles of a trapezoid.*   In Figure 60, refer to trapezoid *KLMN* with $\overline{KL} \parallel \overline{NM}$. Find *x* and *y*.

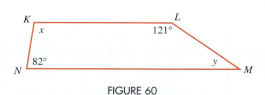

FIGURE 60

Solution

There are two pairs of supplementary angles in this trapezoid: $\angle K$ and $\angle N$, and $\angle L$ and $\angle M$. Since the sum of the measures of supplementary angles is 180°, we have

$$\text{m}(\angle K) + \text{m}(\angle N) = 180° \qquad \text{m}(\angle L) + \text{m}(\angle M) = 180°$$
$$x + 82° = 180° \qquad\qquad 121° + y = 180°$$
$$x = 98° \qquad\qquad\qquad y = 59°$$

Thus, $x = 98°$ and $y = 59°$.

Self Check

Refer to trapezoid *HIJK* below, with $\overline{HI} \parallel \overline{KJ}$. Find *x* and *y*.

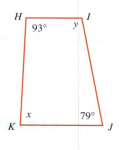

Answers: $x = 87°$, $y = 101°$ ■

If the nonparallel sides of a trapezoid are the same length, it is called an **isosceles trapezoid.** Figure 61 shows isosceles trapezoid *DEFG* with $\overline{DG} \cong \overline{EF}$. In an isosceles trapezoid, *both pairs of base angles are congruent.* In the figure, $\angle D \cong \angle E$ and $\angle G \cong \angle F$.

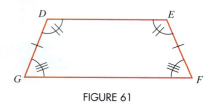

FIGURE 61

EXAMPLE 4   *Cross section of a drainage ditch.*
A cross section of a drainage ditch (Figure 62) is an isosceles trapezoid with $\overline{AB} \parallel \overline{DC}$. Find x and y.

Solution   Since the figure is an isosceles trapezoid, its nonparallel sides have the same length. So $m(\overline{AD})$ and $m(\overline{BC})$ are equal, and $x = 8$ ft.
   Since the base angles of an isosceles trapezoid are congruent, $m(\angle D) = m(\angle C)$. Thus, $y = 120°$.

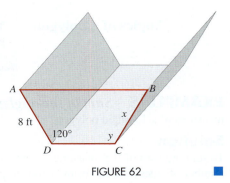

FIGURE 62

The sum of the measures of the angles of a polygon

In Figure 63, a protractor has been used to find the measure of each angle of the quadrilateral. When we add the four angle measures, the result is 360°.

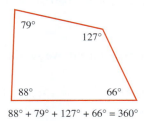

$$88° + 79° + 127° + 66° = 360°$$

FIGURE 63

This illustrates an important fact about quadrilaterals: The sum of the measures of the angles of *any* quadrilateral is 360°. This can be shown using the diagram in Figure 64(a). In the figure, the quadrilateral is divided into two triangles. Since the sum of the angle measures of any triangle is 180°, the sum of the measures of the angles of the quadrilateral is 2 · 180° or 360°.
   A similar approach can be used to find the sum of the measures of the angles of any pentagon or any hexagon. The pentagon in Figure 65(b) is divided into three triangles. The sum of the measures of the angles of the pentagon is 3 · 180° or 540°. The hexagon in Figure 64(c) is divided into four triangles. The sum of the measures of the angles of the hexagon is 4 · 180° or 720°. In general, a polygon with n sides can be divided into $n - 2$ triangles. Therefore, the sum of the angle measures of a polygon can be found by multiplying 180° by $n - 2$.

Quadrilateral

Pentagon

Hexagon

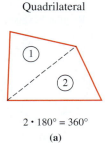

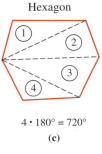

$2 \cdot 180° = 360°$ $3 \cdot 180° = 540°$ $4 \cdot 180° = 720°$

(a)   (b)   (c)

FIGURE 64

Angles of a polygon

The sum S, in degrees, of the measures of the angles of a polygon with n sides is given by the formula

$$S = (n - 2)180°$$

EXAMPLE 5   *Sum of the angles of a polygon.*   Find the sum of the angle measures of a 13-sided polygon.

Solution

To find the sum of the measures of the angles of the polygon, we substitute 13 for the number of sides n in the formula and simplify.

$S = (n - 2)180°$

$S = (13 - 2)180°$ Substitute 13 for n.

$\ \ = (11)180°$ Do the subtraction within the parentheses.

$\ \ = 1{,}980°$

The sum of the measures of the angles of a 13-sided polygon is 1,980°.

Self Check

Find the sum of the angle measures of the polygon shown below.

Answer: 900°   ∎

EXAMPLE 6   *Finding the number of sides of a polygon.*   The sum of the measures of the angles of a polygon is 1,080°. Find the number of sides the polygon has.

Solution

To find the number of sides the polygon has, we substitute 1,080° for S in the formula and then solve for n.

$S = (n - 2)180°$

$\mathbf{1{,}080°} = (n - 2)180°$ Substitute 1,080° for S.

$1{,}080° = 180°n - 360°$ Distribute the multiplication by 180°.

$1{,}080° \mathbf{+ 360°} = 180°n - 360° \mathbf{+ 360°}$ To undo the subtraction of 360°, add 360° to both sides.

$1{,}440° = 180°n$ Do the additions.

$\dfrac{1{,}440°}{\mathbf{180°}} = \dfrac{180°n}{\mathbf{180°}}$ To undo the multiplication by 180°, divide both sides by 180°.

$8 = n$ Do the division: $\frac{1{,}440°}{180°} = 8$.

The polygon has 8 sides. It is an octagon.

Self Check

The sum of the measures of the angles of a polygon is 1,620°. Find the number of sides the polygon has.

Answer: 11   ∎

Finding an angle measure of a regular polygon

Recall that a polygon that has all sides the same length and all angles the same measure is called a *regular polygon*. Two regular polygons are shown in Figure 65.

Regular hexagon   Regular decagon

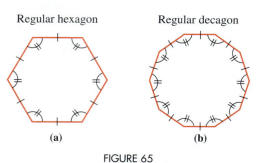

(a)   (b)

FIGURE 65

We can find the measure of one angle of a regular polygon using the following formula.

Angle measure of a regular polygon

The angle measure A, in degrees, of one angle of a regular polygon with n sides is given by the formula

$$A = \frac{(n-2)180°}{n}$$

EXAMPLE 7   *Finding an angle measure of a regular polygon.*   Find the measure of one angle of the polygon in Figure 65(a).

Solution

The regular hexagon shown in Figure 65(a) has six angles with the same measure. To find the measure of one of its angles, we substitute 6 for n in the formula and simplify.

$$A = \frac{(n-2)180°}{n}$$

$$A = \frac{(6-2)180°}{6}$$

$$A = \frac{(4)180°}{6} \quad \text{Do the subtraction within the parentheses.}$$

$$A = \frac{720°}{6} \quad \text{Do the multiplication in the numerator.}$$

$$A = 120° \quad \text{Do the division.}$$

The measure of one angle of a regular hexagon is 120°.

Self Check

Find the measure of one angle of the polygon shown in Figure 65(b).

Answer: 144°

When we know the measure of one angle of a regular polygon, we can use the following formula to find the number of sides the polygon has.

Regular polygons

The number of sides n that a regular polygon has is given by the formula

$$n = \frac{360°}{180° - A}$$

where A is the measure, in degrees, of one angle of the polygon.

EXAMPLE 8   *Determining the number of sides of a regular polygon.*
The measure of each angle of a regular polygon is 150°. How many sides does it have?

Solution

We can find the number of sides of the polygon by substituting 150° for A in the formula for the angle measure of a regular polygon and solving for n.

$$n = \frac{360°}{180° - A}$$

$$n = \frac{360°}{180° - 150°} \quad \text{Substitute 150° for } A.$$

$$= \frac{360°}{30°} \quad \text{Do the subtraction.}$$

$$= 12$$

The polygon has 12 sides.

Self Check

The measure of each angle of a regular polygon is 165°. How many sides does it have?

Answer: 24

Exterior angles of a regular polygon

The pentagon *ABCDE* shown in Figure 66 has five **interior angles** that are highlighted in red. The five angles that are highlighted in blue are called **exterior angles** of the polygon.

Exterior angle of a polygon

> An **exterior angle** of a polygon is an angle that is adjacent and supplementary to an interior angle.

In the figure, $\angle FAB$ is an exterior angle of the polygon, because it is adjacent and supplementary to $\angle BAE$.

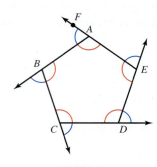

FIGURE 66

Suppose pentagon *ABCDE* in Figure 66 is a regular polygon. To find the measure *A* of one of its five congruent interior angles, we proceed as follows.

$$A = \frac{(n - 2)180°}{n}$$ The formula for the measure of an interior angle of a regular polygon.

$$A = \frac{(5 - 2)180°}{5}$$ Because the regular pentagon has five sides, substitute 5 for *n*.

$$A = \frac{(3)180°}{5}$$

$$= \frac{540°}{5}$$

$$= 108°$$

The measure of each interior angle of the regular pentagon *ABCDE* is 108°. Since each exterior angle of the pentagon is supplementary to an interior angle, each exterior angle has measure $180° - 108° = 72°$. For example, m($\angle BAE$) = 108°, and m($\angle FAB$) = $180° - 108° = 72°$.

Measure of an exterior angle of a regular polygon

> If *A* is the measure of an interior angle and *E* is the measure of an exterior angle of a regular polygon, then
>
> $$E = 180° - A$$

We can derive another formula for the measure of an exterior angle of a regular polygon by replacing *A*, the measure of an interior angle, with $\frac{(n - 2)180°}{n}$ and simplifying.

$$E = 180° - A$$

$$E = 180° - \left[\frac{(n-2)180°}{n}\right]$$

$$E = 180° - \left[\frac{180°n - 360°}{n}\right] \quad \text{Distribute the multiplication by } 180°.$$

$$= 180° - \left[\frac{180°n}{n} - \frac{360°}{n}\right] \quad \begin{array}{l}\text{Within the brackets: Divide each term of the} \\ \text{numerator by } n.\end{array}$$

$$= 180° - \left[180° - \frac{360°}{n}\right] \quad \text{In the first fraction, divide out the common factor of } n.$$

$$= 180° - 180° + \frac{360°}{n}$$

$$= \frac{360°}{n}$$

Measure of an exterior angle of a polygon

> If E is the measure of an exterior angle of a regular polygon with n sides, then
>
> $$E = \frac{360°}{n}$$

Just as it is useful to know the sum of the measures of the interior angles of a regular polygon, it is also helpful to know the sum of the measures of its exterior angles. Since there are n angles in a regular polygon with n sides, the sum S of the measures of the exterior angles, one at each vertex, of a regular polygon is the product of the number of sides n of the polygon and the measure E of one exterior angle.

$$S = n \cdot E$$

$$S = n \cdot \frac{360°}{n} \quad \text{Replace } E \text{ with } \tfrac{360°}{n}.$$

$$S = 360°$$

Sum of the exterior angles of a regular polygon

> The sum of the measures of the exterior angles of a regular polygon is 360°.

STUDY SET   Section 5

VOCABULARY   *Fill in the blanks.*

1. A _____ is a polygon with four sides.

2. A _____ is a quadrilateral with opposite sides parallel.

3. A _____ is a quadrilateral with four right angles.

4. A rectangle with all sides of equal length is a _____.

5. A _____ is a parallelogram with four sides of equal length.

6. A _____ has two sides that are parallel and two sides that are not parallel. The parallel sides of a trapezoid are called _____. The legs of an _____ trapezoid have the same length.

7. A segment that joins two nonconsecutive vertices of a polygon is called a _____ of the polygon.

8. A _____ polygon has sides that are all the same length and angles that are all the same measure.

CONCEPTS

9. Refer to the polygon in Illustration 1.
  a. How many vertices does it have? List them.
  b. How many sides does it have? List them.
  c. How many diagonals does it have? List them.
  d. Tell which of the following are acceptable ways of naming the polygon.

quadrilateral *ABCD*
quadrilateral *CDBA*
quadrilateral *ACBD*
quadrilateral *BADC*

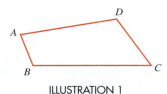

ILLUSTRATION 1

10. Classify each quadrilateral as a rectangle, a square, a rhombus, or a trapezoid.

  a.
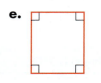

  b.

  c.

  d.
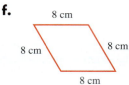

  e.
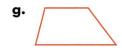

  f.

  g.
  h.

11. Draw an example of each type of quadrilateral.
  a. rhombus   **b.** parallelogram
  c. trapezoid   **d.** square
  e. rectangle   **f.** isosceles trapezoid

12. A parallelogram is shown in Illustration 2. Fill in the blanks.
  a. $\overline{ST} \parallel$ _____   **b.** $\overline{SV}$ ___ $\overline{TU}$

ILLUSTRATION 2

13. Refer to the rectangle in Illustration 3.
  a. How many right angles does the rectangle have? List them.
  b. Which sides are parallel?
  c. Which sides are of equal length?
  d. Copy the figure and draw the diagonals. Call the point where the diagonals intersect point *X*. How many diagonals does the figure have? List them.

ILLUSTRATION 3

14. Fill in the blanks. In any rectangle:
  a. All four angles are _____ angles.
  b. Opposite sides are _____.
  c. Opposite sides have equal _____.
  d. The diagonals have equal _____.
  e. The diagonals intersect at their _____.

15. Refer to Illustration 4.
  a. What is m($\overline{CD}$)?   **b.** What is m($\overline{AD}$)?

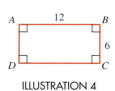

ILLUSTRATION 4

16. Rectangle *ABCD* is shown in Illustration 5.
  a. What is m($\overline{AX}$)?
  b. What is m($\overline{AC}$)?
  c. What is m($\overline{BD}$)?

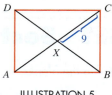

ILLUSTRATION 5

17. See Illustration 6, where $\overline{TR} \parallel \overline{DF}$, $\overline{DT} \parallel \overline{FR}$, and m($\angle D$) = 90°. What type of quadrilateral is *DTRF*?

ILLUSTRATION 6

18. Refer to the parallelogram shown in Illustration 7. If m($\overline{GI}$) = 4 and m($\overline{HJ}$) = 4, what type of figure is quadrilateral *GHIJ?*

ILLUSTRATION 7

19. a. Is every rectangle a square?
   b. Is every square a rectangle?
   c. Is every parallelogram a rectangle?
   d. Is every rectangle a parallelogram?
   e. Is every rhombus a square?
   f. If every square a rhombus?

20. Trapezoid *WXYZ* is shown in Illustration 8. Which sides are parallel?

ILLUSTRATION 8

21. Trapezoid *JKLM* is shown in Illustration 9.
   a. What type of trapezoid is this?
   b. Which angles are the lower base angles?
   c. Which angles are the upper base angles?
   d. Fill in the blanks:

   m($\angle J$) = _____

   m($\angle K$) = _____

   m($\overline{JK}$) = _____

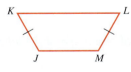

ILLUSTRATION 9

22. Find the sum of the measures of the angles of the hexagon in Illustration 10.

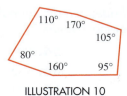

ILLUSTRATION 10

23. Refer to Illustration 11.
   a. How many sides does the polygon have?
   b. If you draw all of the diagonals from one vertex, how many triangles make up the polygon?
   c. What is the difference between the number of sides and the number of triangles?
   d. What formula can be used to find the sum of the angle measures of any polygon?

ILLUSTRATION 11

24. Refer to the regular hexagon in Illustration 12.
   a. Fill in: $\angle BCG$ is called an _____ angle of the hexagon.
   b. Fill in: $\angle BCG$ and $\angle$_____ are adjacent and supplementary.
   c. What is m($\angle BCD$)? What is m($\angle BCG$)?
   d. What is the sum of the exterior angles of the regular hexagon?

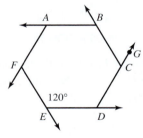

ILLUSTRATION 12

NOTATION

25. What do the tick marks in Illustration 13 indicate?

ILLUSTRATION 13

26. Rectangle *ABCD* is shown in Illustration 14. What do the tick marks indicate about point *X?*

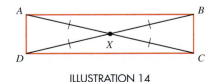

ILLUSTRATION 14

27. a. In the formula $S = (n - 2)180°$, what does S represent? What does n represent?

  b. In the formula $A = \dfrac{(n - 2)180°}{n}$, what does A represent? What does n represent?

  c. In the formula $E = 180° - A$, what does E represent? What does A represent?

  d. In the formula $E = \dfrac{360°}{n}$, what does E represent? What does n represent?

28. Suppose $n = 12$. What is $(n - 2)180°$?

PRACTICE

29. Refer to rectangle $ABCD$ in Illustration 15.
  a. Find m($\angle 1$).   **b.** Find m($\angle 3$).
  c. Find m($\angle 2$).
  d. If m($\overline{AC}$) is 8 cm, find m($\overline{BD}$).
  e. Find m($\overline{PD}$).

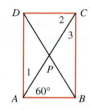

ILLUSTRATION 15

30. Refer to rectangle $EFGH$ in Illustration 16.
  a. Find m($\overline{FH}$).   **b.** Find m($\overline{EI}$).
  c. Find m($\overline{EG}$).

ILLUSTRATION 16

31. Refer to the trapezoid in Illustration 17.
  a. Find x.   **b.** Find y.

ILLUSTRATION 17

32. Refer to trapezoid $MNOP$ in Illustration 18.
  a. Find m($\angle O$).   **b.** Find m($\angle M$).

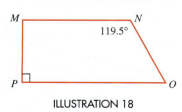

ILLUSTRATION 18

33. Refer to the isosceles trapezoid in Illustration 19.
  a. Find m($\overline{BC}$).   **b.** Find x.
  c. Find y.   **d.** Find z.

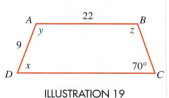

ILLUSTRATION 19

34. Refer to the trapezoid in Illustration 20.
  a. Find m($\angle T$).   **b.** Find m($\angle R$).
  c. Find m($\angle S$).

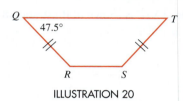

ILLUSTRATION 20

Use the formula discussed in this section to find the sum of the angle measures of each polygon.

35. an octagon   **36.** a decagon
37. a dodecagon   **38.** a nonagon
39. a 16-sided polygon **40.** a 20-sided polygon

Find the number of sides a polygon has if the sum of its angle measures is the given number.

41. $2,160°$ **42.** $3,600°$
43. $900°$ **44.** $720°$
45. $1,440°$ **46.** $1,800°$

Find the measure of one angle of the following regular polygons.

47. an octagon   **48.** a nonagon
49. a pentagon   **50.** a dodecagon
51. a 15-sided polygon **52.** a 100-sided polygon

Find the number of sides of a regular polygon if one of its angles has the following measure.

53. $120°$ **54.** $90°$
55. $162°$ **56.** $160°$
57. $172.8°$ **58.** $179.64°$

APPLICATIONS

59. QUADRILATERALS IN EVERYDAY LIFE What quadrilateral shape do you see in each of the following objects?

a. Podium (upper portion) **b.** Checkerboard

c. Dollar bill **d.** Swimming fin

e. Camper shell window

60. FLOWCHART A flowchart shows a sequence of steps to be performed by a computer to solve a given problem. When designing a flowchart, the programmer uses a set of standardized symbols to represent various operations to be performed by the computer. Locate a rectangle, a rhombus, and a parallelogram in the flow chart shown in Illustration 21.

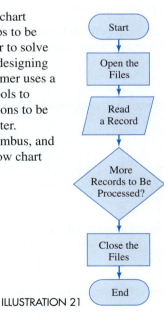

ILLUSTRATION 21

61. MAKING A FRAME After gluing and nailing the pieces of a picture frame together, it didn't look right to a frame maker. See Illustration 22. How can she use a tape measure to make sure the corners are 90° (right) angles?

ILLUSTRATION 22

62. BASEBALL Refer to Illustration 23. Find the sum of the measures of the angles of home plate.

ILLUSTRATION 23

63. TOOLS The utility knife blade shown in Illustration 24 has the shape of an isosceles trapezoid. Find *x*, *y*, and *z*.

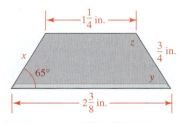

ILLUSTRATION 24

64. TRAFFIC SIGNS The stop sign shown in Illustration 25 is a regular polygon. What is the measure of one of its angles?

ILLUSTRATION 25

WRITING

65. Explain why a square is a rectangle.

66. Explain why a trapezoid is not a parallelogram.

67. Consider a regular polygon with 8 sides and the measure of one of its angles. Consider a regular polygon with 9 sides and the measure of one of its angles. Without using a formula, can you determine which angle measure would be larger? Explain your thinking.

68. A decagon is a polygon with ten sides. What could you call a polygon with one hundred sides? With one thousand sides? With one million sides?

6   *Perimeters and Areas of Polygons*

In this section, you will learn how to find

• Perimeters of polygons • Areas of polygons • Areas of figures that are combinations of polygons

INTRODUCTION. In this section, we will discuss how to find perimeters and areas of polygons. Finding perimeters is important when estimating the cost of fencing or estimating the cost of woodwork in a house. Finding areas is important when calculating the cost of carpeting, the cost of painting a house, or the cost of fertilizing a yard.

Perimeters of polygons

The **perimeter** of a polygon is the distance around it. To find the perimeter P of a polygon, we simply add the lengths of its sides.

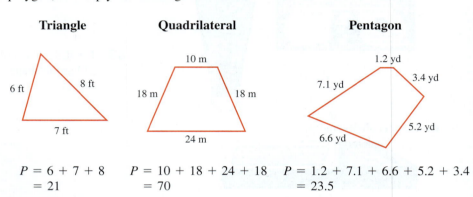

Triangle	Quadrilateral	Pentagon

$$P = 6 + 7 + 8 \qquad P = 10 + 18 + 24 + 18 \qquad P = 1.2 + 7.1 + 6.6 + 5.2 + 3.4$$
$$= 21 \qquad\qquad\qquad = 70 \qquad\qquad\qquad\qquad = 23.5$$

The perimeter is 21 ft.   The perimeter is 70 m.   The perimeter is 23.5 yd.

For some polygons, such as a square and a rectangle, we can simplify the computations by using a perimeter formula. Since a square has four sides of equal length s, its perimeter P is $s + s + s + s$, or $4s$.

Perimeter of a square

If a square has a side of length s, its perimeter P is given by the formula

$$P = 4s$$

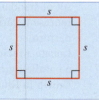

EXAMPLE 1   *Perimeter of a square.*   Find the perimeter of a square whose sides are 7.5 meters long.

Solution

Since the perimeter of a square is given by the formula $P = 4s$, we substitute 7.5 for s and simplify.

$$P = 4s$$
$$P = 4(7.5)$$
$$P = 30$$

The perimeter is 30 meters.

Self Check

A Scrabble game board has a square shape with sides of length 38.5 cm. Find the perimeter of the game board.

Answer: 154 cm

Since a rectangle has two lengths *l* and two widths *w*, its perimeter *P* is *l* + *w* + *l* + *w*, or 2*l* + 2*w*.

Perimeter of a rectangle

If a rectangle has length *l* and width *w*, its perimeter *P* is given by the formula

$$P = 2l + 2w$$

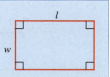

 COMMENT   When finding the perimeter of a polygon, the lengths of the sides must be expresed in the same units.

EXAMPLE 2   *Converting units.*   Find the perimeter of the rectangle in Figure 67, in meters.

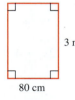

3 m

80 cm

FIGURE 67

Solution
Before we can find the perimeter of the rectangle, we must express the length and width in terms of the same units. Since 1 meter = 100 centimeters, we can convert 80 centimeters to meters by multiplying 80 centimeters by the unit conversion factor $\frac{1\,m}{100\,cm}$.

$$80\ cm = 80\ cm \cdot \frac{1\ m}{100\ cm} \qquad \text{Multiply by 1:} \frac{1\ m}{100\ cm} = 1.$$

$$= \frac{\overset{1}{\cancel{80\ cm}}}{1} \cdot \frac{1\ m}{\underset{1}{\cancel{100\ cm}}} \qquad \text{Write 80 cm as a fraction: } 80\ cm = \frac{80\ cm}{1}. \text{ The units of centimeters divide out.}$$

$$= \frac{80}{100}\ m$$

$$= 0.8\ m \qquad \text{Divide by 100 by moving the understood decimal point in 80 two places to the left.}$$

The width of the rectangle is 0.8 m. We can now substitute 3 for *l* and 0.8 for *w* in the formula for the perimeter of rectangle to get

$$P = 2\boldsymbol{l} + 2\boldsymbol{w}$$

$$P = 2(\mathbf{3}) + 2(\mathbf{0.8})$$

$$= 6 + 1.6$$

$$= 7.6$$

The perimeter is 7.6 meters.

Self Check

Find the perimeter of the triangle below, in inches.

14 in.   12 in.

2 ft

Answer: 50 in.   ■

EXAMPLE 3   *Structural engineering.*   The truss shown in Figure 68 is made up of three wooden components that form an isosceles triangle. The length of the base is 4 feet less than twice the length of one of the sides. If 76 linear feet of lumber were used to make the truss, how long is each component of the truss?

s   *s*

2*s* − 4 ft

FIGURE 68

Analyze the problem
- The truss is in the shape of an isosceles triangle.
- The length of the base is 4 feet less than twice the length of a side.
- The perimeter of the truss is 76 feet.
- We are to find the length of each component of the truss.

Form an equation   Since the length of the base is related to the length of a side, we begin by letting $s =$ the length of a side component (in feet). Then we translate the words *4 feet less than twice the side,* to get an algebraic expression to represent the length of the base.

$$2s - 4 = \text{the length of the base (in feet)}$$

Because 76 linear feet of lumber were used to make the triangular-shaped truss,

The length of the base of the truss	plus	the length of one side	plus	the length of the other side	equals	the perimeter of the truss.
$2s - 4$	$+$	s	$+$	s	$=$	76

Solve the equation

$$2s - 4 + s + s = 76$$
$$4s - 4 = 76 \qquad \text{Combine like terms: } 2s + s + s = 4s.$$
$$4s = 80 \qquad \text{To undo the subtraction of 4, add 4 to both sides.}$$
$$\frac{4s}{4} = \frac{80}{4} \qquad \text{To undo the multiplication by 4, divide both sides by 4.}$$
$$s = 20 \qquad \text{Do the divisions.}$$

To find the length of the base, we evaluate the expression $2s - 4$ for $s = 20$.

$$2s - 4 = 2(20) - 4$$
$$= 36$$

State the conclusion   The length of the base component is 36 ft, and the length of each side component is 20 ft.

Check the result   If we add the lengths of the components of the truss, we get 36 ft + 20 ft + 20 ft = 76 ft. The answers check. ■

Accent on Technology: **Perimeters of figures that are combinations of polygons**

See Figure 69. To find the perimeter, we need to know the values of x and y. Since the figure is a combination of two rectangles, we can use a calculator to see that

$$x = 20.25 - 10.17 \qquad \text{and} \qquad y = 12.5 - 4.75$$
$$x = 10.08 \text{ cm} \qquad\qquad y = 7.75 \text{ cm}$$

The perimeter P of the figure is

$$P = 20.25 + 12.5 + 10.17 + 4.75 + x + y$$
$$P = 20.25 + 12.5 + 10.17 + 4.75 + 10.08 + 7.75$$

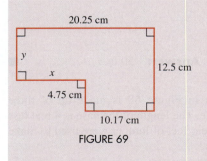

FIGURE 69

We can use a calculator to evaluate the expression on the right-hand side by entering these numbers and pressing these keys.

Keystrokes   20.25 $+$ 12.5 $+$ 10.17 $+$ 4.75 $+$ 10.08 $+$ 7.75 $=$

$$\boxed{65.5}$$

The perimeter is 65.5 centimeters.

Areas of polygons

The **area** of a polygon is the measure of the amount of surface it encloses. Area is measured in square units, such as square inches or square centimeters. See Figure 70.

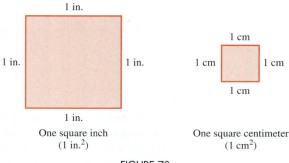

FIGURE 70

In everyday life, we often use areas. For example,

- To carpet a room, we buy square yards.
- A can of paint will cover a certain number of square feet.
- To measure vast amounts of land, we often use square miles.
- We buy house roofing by the "square." One square is 100 square feet.

The rectangle shown in Figure 71 has a length of 10 centimeters and a width of 3 centimeters. If we divide the rectangle into squares as shown in the figure, each square represents an area of 1 square centimeter—a surface enclosed by a square measuring 1 centimeter on each side. Because there are 3 rows with 10 squares in each row, there are 30 squares. Since the rectangle encloses a surface area of 30 squares, its area is 30 square centimeters, which can be written as 30 cm^2.

This example illustrates that to find the area of a rectangle, we multiply its length by its width.

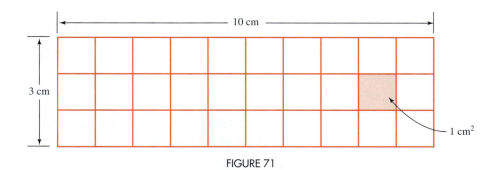

FIGURE 71

 COMMENT   Do not confuse the concepts of perimeter and area. Perimeter is the distance around a polygon. It is measured in linear units, such as centimeters, feet, or miles. Area is a measure of the surface enclosed within a polygon. It is measured in square units, such as square centimeters, square feet, or square miles.

In practice, we do not find areas of polygons by counting squares. Instead, we use formulas to find areas of geometric figures. We have seen that the area of a rectangle is the product its length and width. This fact can be used to derive the area formula for a parallelogram.

Figure 72 shows how a parallelogram with base b and height h can be transformed into a rectangle with length b and width h. Since the area of the rectangle is bh, the area of the parallelogram is also bh. Therefore, the area A of the parallelogram is given by the formula $A = bh$.

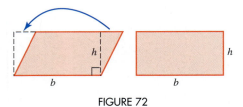

FIGURE 72

To derive the formula for the area of a triangle, we use the fact that the area of a parallelogram is the product of the length of its base and its height. Figure 73 shows how a triangle with base b and height h can, with the addition of an identical triangle, be transformed into a parallelogram with base b and height h. Since the area of the parallelogram is bh, the area of the original triangle must be one-half of that. Therefore, the area A of the original triangle is given by the formula $A = \frac{1}{2}bh$.

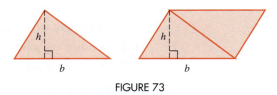

FIGURE 73

We can derive the formula for a trapezoid using the formula for the area of a parallelogram. Figure 74 shows how a trapezoid with bases b_1 and b_2 and height h can, with the addition of an identical trapezoid, be transformed into a parallelogram with base $(b_1 + b_2)$ and height h. Since the area of the parallelogram is $(b_1 + b_2)h$, the area of the original trapezoid must be one-half of that. Therefore, the area A of the original trapezoid is given by the formula $A = \frac{1}{2}(b_1 + b_2)h$.

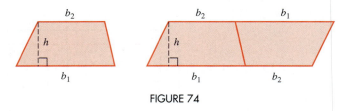

FIGURE 74

The formulas for finding the area of several types of polygons are summarized in Table 1 on the next page.

EXAMPLE 4   *Area of a square.*   Find the area of the square in Figure 75.

Solution
We can see that the length of one side of the square is 15 centimeters. We can find its area by using the formula $A = s^2$ and substituting 15 for s.

$A = s^2$

$A = (\mathbf{15\ cm})^2$ Substitute 15 for s.

$A = 225\ cm^2$ Evaluate the exponential expression: $15 \cdot 15 = 225$.

The area of the square is 225 cm².

FIGURE 75

Self Check
Find the area of the square shown below.

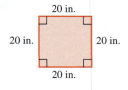

Answer: 400 in.²

Figure	Name	Formula for area
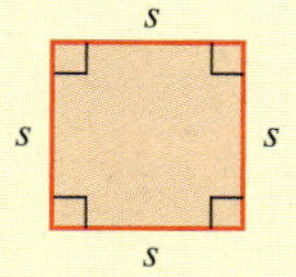	Square	$A = s^2$, where s is the length of one side.
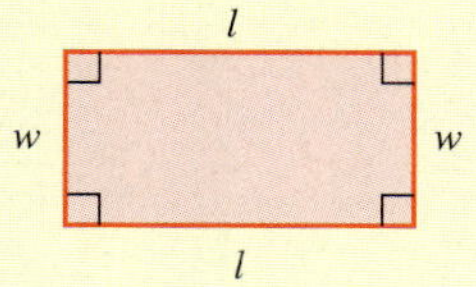	Rectangle	$A = lw$, where l is the length and w is the width.
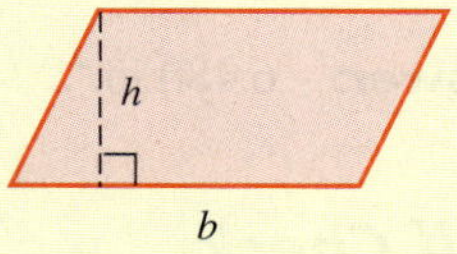	Parallelogram	$A = bh$, where b is the length of the base and h is the height. (A height is always perpendicular to the base.)
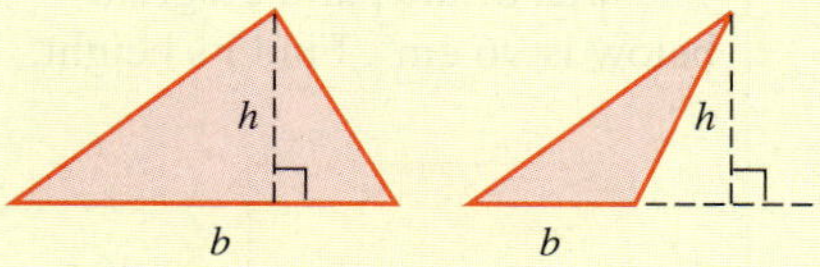	Triangle	$A = \frac{1}{2}bh$, where b is the length of the base and h is the height. The segment perpendicular to the base and representing the height is called an **altitude**.
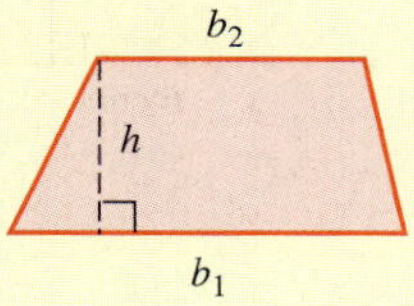	Trapezoid	$A = \frac{1}{2}h(b_1 + b_2)$, where h is the height of the trapezoid and b_1 and b_2 represent the lengths of the bases.

TABLE 1

EXAMPLE 5   *Number of square feet in 1 square yard.*
Find the number of square feet in 1 square yard. (See Figure 76.)

Solution
Since 3 feet = 1 yard, each side of 1 square yard is 3 feet long.

$$1 \text{ yd}^2 = (\textbf{1 yd})^2$$
$$= (\textbf{3 ft})^2 \quad \text{Substitute 3 feet for 1 yard.}$$
$$= 9 \text{ ft}^2 \quad (3 \text{ ft})^2 = (3 \text{ ft})(3 \text{ ft}) = 9 \text{ ft}^2.$$

There are 9 square feet in 1 square yard.

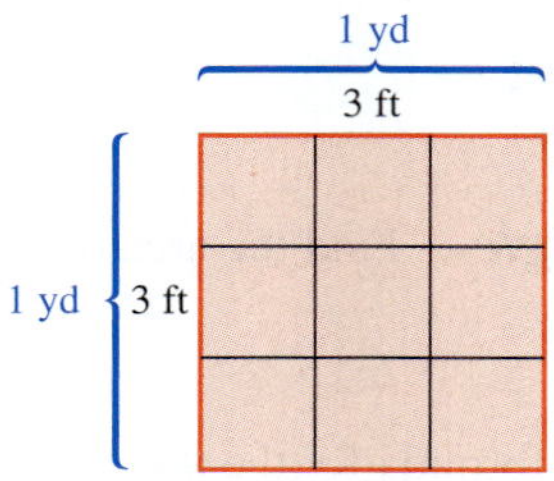

FIGURE 76

Self Check
Find the number of square centimeters in 1 square meter.

Answer: 10,000 cm²

EXAMPLE 6   *Women's sports.*
Field hockey is a team sport in which players use sticks to try to hit a ball into their opponents' goal. Find the area of the rectangular field shown in Figure 77. Give the answer in square feet.

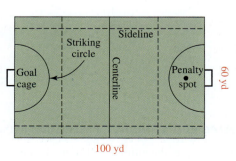

FIGURE 77

Self Check
A regulation size ping-pong table is 9 feet long and 5 feet wide. Find its area in square inches.

Solution

To find the area in square yards, we substitute 100 for l and 60 for w in the formula for the area of a rectangle, and simplify.

$$A = \boldsymbol{lw}$$

$$A = \boldsymbol{100}(\boldsymbol{60})$$

$$= 6{,}000$$

The area is 6,000 square yards. Since there are 9 square feet per square yard, we can convert this number to square feet by multiplying 6,000 square yards by $\frac{9 \text{ ft}^2}{1 \text{ yd}^2}$.

$$6{,}000 \text{ yd}^2 = 6{,}000 \text{ yd}^2 \cdot \frac{9 \text{ ft}^2}{1 \text{ yd}^2} \qquad \text{\color{red}{Multiply by the unit conversion factor: } } \tfrac{9 \text{ ft}^2}{1 \text{ yd}^2}.$$

$$= 6{,}000 \cdot 9 \text{ ft}^2 \qquad \text{\color{red}{The units of square yards divide out.}}$$

$$= 54{,}000 \text{ ft}^2 \qquad \text{\color{red}{Multiply: } } 6{,}000 \cdot 9 = 54{,}000.$$

The area of the field is 54,000 ft^2.

Answer: 6,480 in.^{2} ■

EXAMPLE 7   *Finding the height of a parallelogram.*
The area of the parallelogram shown in Figure 78 is 360 ft^2. Find the height.

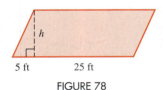

FIGURE 78

Self Check

The area of the parallelogram below is 96 cm^2. Find its height.

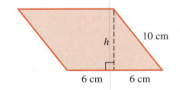

Solution

The length of the base of the parallelogram is

$$5 \text{ feet} + 25 \text{ feet} = 30 \text{ feet}$$

We let h = the height of the parallelogram. Then we substitute 360 for A and 30 for b in the formula for the area of a parallelogram and solve for h.

$$\boldsymbol{A} = \boldsymbol{bh}$$

$$\boldsymbol{360} = \boldsymbol{30}h$$

$$\frac{\boldsymbol{360}}{\boldsymbol{30}} = \frac{\boldsymbol{30}h}{\boldsymbol{30}} \qquad \text{\color{red}{To undo the multiplication by 30, divide both sides by 30.}}$$

$$12 = h$$

The height of the parallelogram is 12 feet.

Answer: 8 cm   ■

EXAMPLE 8   *Area of a triangle.*   Find the area of the triangle in Figure 79.

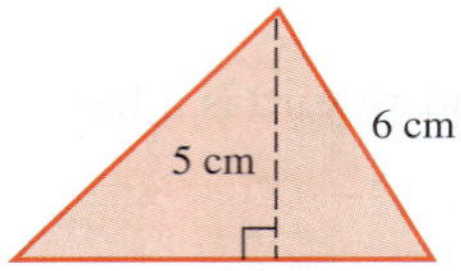

Self Check

Find the area of the triangle below.

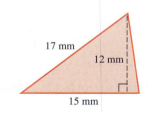

Solution

We substitute 8 for b and 5 for h in the formula for the area of a triangle, and simplify. (The side having length 6 cm is additional information that is not used to find the area.)

$$A = \frac{1}{2}\boldsymbol{bh}$$

$$A = \frac{1}{2}(\boldsymbol{8})(\boldsymbol{5}) \qquad \text{\color{red}{The length of the base is 8 cm. The height is 5 cm.}}$$

$$= 4(5) \qquad \text{\color{red}{Do the multiplication: } } \tfrac{1}{2}(8) = 4.$$

$$= 20$$

The area of the triangle is 20 cm^2.

Answer: 90 mm^2 ■

FIGURE 79

EXAMPLE 9   *Area of a triangle.*   Find the area of the triangle in Figure 80.

Solution   In this case, the altitude falls outside the triangle.

$$A = \frac{1}{2}\textbf{bh}$$

$$A = \frac{1}{2}(\textbf{9})(\textbf{13})$$ Substitute 9 for *b* and 13 for *h*.

$$= \frac{1}{2}\left(\frac{9}{1}\right)\left(\frac{13}{1}\right)$$ Write 9 as $\frac{9}{1}$ and 13 as $\frac{13}{1}$.

$$= \frac{117}{2}$$ Multiply the fractions.

$$= 58.5$$ Do the division.

The area of the triangle is 58.5 cm².

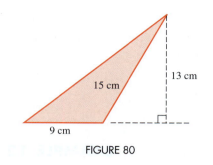

FIGURE 80

EXAMPLE 10   *Area of a trapezoid.*   Find the area of the trapezoid in Figure 81.

Solution
In this example, $b_1 = 10$ and $b_2 = 6$. It is incorrect to say that $h = 1$, because the height of 1 foot must be expressed as 12 inches to be consistent with the units of the bases. Thus, we substitute 10 for b_1, 6 for b_2, and 12 for h in the formula for finding the area of a trapezoid and simplify.

$$A = \frac{1}{2}\textbf{h}(\textbf{b}_1 + \textbf{b}_2)$$

$$A = \frac{1}{2}(\textbf{12})(\textbf{10} + \textbf{6})$$ The length of the lower base is 10 in. The length of the upper base is 6 in. The height is 12 in.

$$= \frac{1}{2}(12)(16)$$ Do the addition within the parentheses.

$$= 6(16)$$ Do the multiplication: $\frac{1}{2}(12) = 6$.

$$= 96$$

The area of the trapezoid is 96 in.²

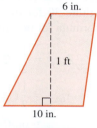

FIGURE 81

Self Check
Find the area of the trapezoid below.

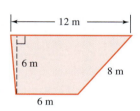

Answer: 54 m²

Areas of figures that are combinations of polygons

EXAMPLE 11   *Carpeting a room.*   A living room/dining room area has the floor plan shown in Figure 82. If carpet costs $29 per square yard, including pad and installation, how much will it cost to carpet the room? (Assume no waste.)

Solution   First we must find the total area of the living room and the dining room:

$$A_{\text{total}} = A_{\text{living room}} + A_{\text{dining room}}$$

Since $\overline{CF}$ divides the space into two rectangles, the areas of the living room and the dining room are found by multiplying their respective lengths and widths. Therefore, the area

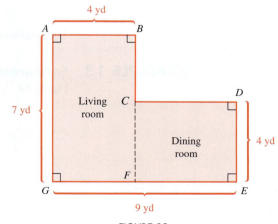

FIGURE 82

of the living room is $7 \text{ yd} \cdot 4 \text{ yd} = 28 \text{ yd}^2$, and the area of the dining room is $5 \text{ yd} \cdot 4 \text{ yd} = 20 \text{ yd}^2$. The total area to be carpeted is the sum of these two areas.

$$A_{\text{total}} = A_{\text{living room}} + A_{\text{dining room}}$$
$$A_{\text{total}} = \mathbf{28 \ yd^2} + \mathbf{20 \ yd^2}$$
$$= 48 \text{ yd}^2$$

At \$29 per square yard, the cost to carpet the room will be $48 \cdot \$29$, or \$1,392. ∎

EXAMPLE 12 *Combinations of polygons.* Find the area of one side of the tent in Figure 83.

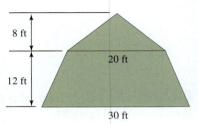

8 ft

20 ft

12 ft

30 ft

FIGURE 83

Solution Each side is a combination of a trapezoid and a triangle. Since the bases of the trapezoid are 30 feet and 20 feet and the height is 12 feet, we substitute 30 for b_1, 20 for b_2, and 12 for h in the formula for the area of a trapezoid.

$$A_{\text{trap.}} = \frac{1}{2}\boldsymbol{h(b_1 + b_2)}$$

$$A_{\text{trap.}} = \frac{1}{2}\mathbf{(12)(30 + 20)}$$

$$= 6(50) \qquad \textcolor{red}{\text{First, do the addition within the parentheses. Then do the multiplication: } \frac{1}{2}(12) = 6.}$$

$$= 300$$

The area of the trapezoid is 300 ft^2.

Since the triangle has a base of 20 feet and a height of 8 feet, we substitute 20 for b and 8 for h in the formula for the area of a triangle.

$$A_{\text{triangle}} = \frac{1}{2}\boldsymbol{bh}$$

$$A_{\text{triangle}} = \frac{1}{2}\mathbf{(20)(8)}$$

$$= 80 \qquad \textcolor{red}{\text{Do the multiplications working from left to right: } \frac{1}{2}(20) = 10 \text{ and then } 10(8) = 80.}$$

The area of the triangle is 80 ft^2.

The total area of one side of the tent is

$$A_{\text{total}} = A_{\text{trap.}} + A_{\text{triangle}}$$
$$A_{\text{total}} = \mathbf{300 \ ft^2} + \mathbf{80 \ ft^2}$$
$$= 380 \text{ ft}^2$$

The total area is 380 ft^2. ∎

EXAMPLE 13 *Subtracting out unwanted area.* Find the area of the shaded region shown in Figure 84.

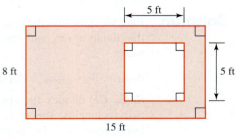

5 ft

8 ft

5 ft

15 ft

FIGURE 84

Solution

The area of the shaded region can be found by calculating the area of the rectangle and then subtracting the area of the square from it.

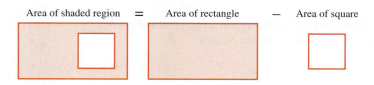

Area of shaded region   =   Area of rectangle   —   Area of square

$$A_{\text{shaded}} = \mathbf{l}\mathbf{w} - \mathbf{s}^2$$ The formula for the area of a rectangle is $A = lw$, and the formula for the area of a square is $A = s^2$.

$$= \mathbf{15}(\mathbf{8}) - \mathbf{5}^2$$ To find the area of the rectangle, substitute 15 for the length l and 8 for the width w. To find the area of the square, substitute 5 for the length s of a side.

$$= 120 - 25$$

$$= 95$$

The area of the shaded region is 95 ft^2.

STUDY SET   Section 6

VOCABULARY   *Fill in the blanks.*

1. The distance around a polygon is called the _____.

2. The _____ of a polygon is measured in linear units such as inches, feet, and miles.

3. The measure of the surface enclosed by a polygon is called its _____.

4. If each side of a square measures 1 foot, the area enclosed by the square is 1 _____ foot.

5. The _____ of a polygon is measured in square units.

6. The segment that represents the height of a triangle is called an _____.

CONCEPTS

7. Illustration 1 shows a kitchen floor that is covered with 1-foot-square tiles. What is the area of the floor?

ILLUSTRATION 1

8. Tell which concept applies, perimeter or area.
 a. The length of a walk around New York's Central Park
 b. The amount of land in Yellowstone National Park
 c. The amount of fence needed to enclose a playground
 d. The amount of office floor space in a building

9. For each figure below, draw the altitude to the base b.

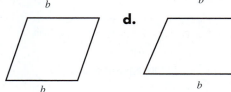

10. For each figure below, label the base b for the given altitude.

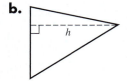

c. 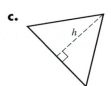   **d.**

11. The shaded figure in Illustration 2 is a combination of what two types of geometric figures?

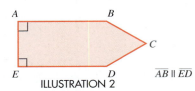

$\overline{AB} \parallel \overline{ED}$

ILLUSTRATION 2

12. Explain how you would find the area of the shaded figure in Illustration 3.

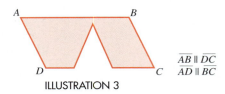

$\overline{AB} \parallel \overline{DC}$
$\overline{AD} \parallel \overline{BC}$

ILLUSTRATION 3

Sketch and label each of the figures.

13. Two different rectangles, each having a perimeter of 40 in.

14. Two different rectangles, each having an area of 40 in.^{2}.

15. A square with an area of 25 m^2.

16. A square with a perimeter of 20 m.

17. A parallelogram with an area of 15 yd^2.

18. A triangle with an area of 20 ft^2.

19. A figure consisting of a combination of two rectangles whose total area is 80 ft^2.

20. A figure consisting of a combination of a rectangle and a square whose total area is 164 ft^2.

21. Refer to Illustration 4. What must be done before we can use the formula to find the area of this rectangle?

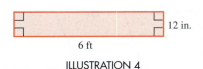

12 in.

6 ft

ILLUSTRATION 4

22. A student expressed the area of the square in Illustration 5 as 25^2 ft. Explain his error.

5 ft

5 ft

ILLUSTRATION 5

23. The lengths of the sides of the polygons are represented by algebraic expressions. In each case, the units are feet. Find the perimeter of the polygon.

a.

$x + 1$

b.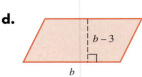

x

$x + 3$

c.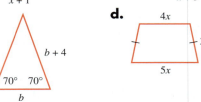

$b + 4$

70°   70°

b

d.

4x

3x

5x

24. The dimensions of the polygons below are represented by algebraic expressions. In each case, the units are meters. Find the area of the polygon.

a.

2x

b.

x

$x + 3$

c.

h

$h + 4$

d.

$b - 3$

b

25. How many square inches are in 1 square foot?

26. How many square inches are in 1 square yard?

NOTATION   *Fill in the blanks.*

27. The formula for the perimeter of a square is _____.

28. The formula for the perimeter of a rectangle is _____.

29. The symbol 1 in.^{2} means one _____.

30. One square meter is expressed as _____.

31. The formula for the area of a square is _____.

32. The formula for the area of a rectangle is _____.

33. The formula $A = \frac{1}{2}bh$ gives the area of a _____.

34. The formula $A = \frac{1}{2}h(b_1 + b_2)$ gives the area of a _____.

35. The formula for the area of a parallogram is _____.

36. In Illustration 6, the symbol ⌐ indicates that the dashed line segment, called an *altitude*, is _____ to the base.

ILLUSTRATION 6

PRACTICE   *Find the perimeter of each figure.*

37.

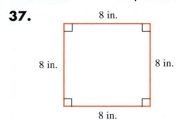

38.

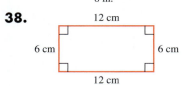

39.

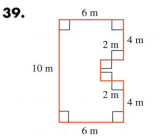

40.

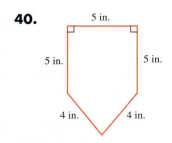

Find x and y. Then find the perimeter of the figure.

41.

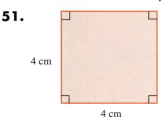

42.

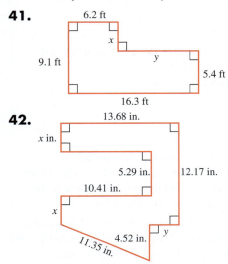

Solve each problem.

43. Find the perimeter of an isosceles triangle with a base of length 21 centimeters and sides of length 32 centimeters.

44. The perimeter of an isosceles triangle is 80 meters. If the length of one side is 22 meters, how long is the base?

45. The perimeter of a square is 35 yards. How long is a side of the square?

46. The perimeter of an equilateral triangle is 85 feet. Find the length of each side.

47. An isosceles triangle with congruent sides of length 49.3 inches has a perimeter of 121.7 inches. Find the length of the base.

48. The perimeter of a rectangle is 80 millimeters. The length is 8 mm longer than the width. Find its length and width.

49. The perimeter of an isosceles trapezoid is 35 meters. The upper base is 5 meters shorter than the lower base. Each leg is 10 meters shorter than the lower base. How long is each side of the trapezoid?

50. The perimeter of an isosceles triangle is 94 feet. Each of the congruent sides is 2 feet more than four times as long as the base. Find the length of each side of the triangle.

Find the area of the shaded part of each figure.

51.

52.

53.

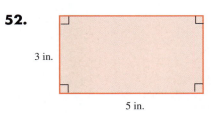

$\overline{AB} \parallel \overline{DC}$
$\overline{BC} \parallel \overline{AD}$

54.

$\overline{AB} \parallel \overline{DC}$
$\overline{AD} \parallel \overline{BC}$

55.

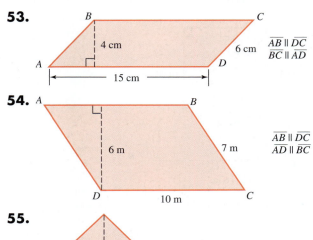

56.

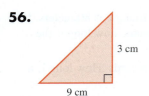

3 cm

9 cm

57.

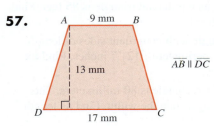

A 9 mm B

$\overline{AB} \parallel \overline{DC}$

13 mm

D 17 mm C

58.

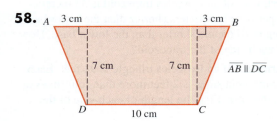

A 3 cm   3 cm B

7 cm   7 cm $\overline{AB} \parallel \overline{DC}$

D 10 cm C

59.

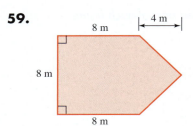

8 m   4 m

8 m

8 m

60.

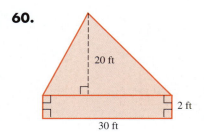

20 ft

2 ft

30 ft

61.

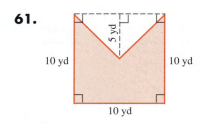

5 yd

10 yd   10 yd

10 yd

62.

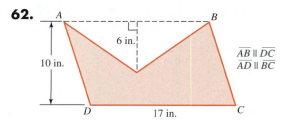

A B

6 in. $\overline{AB} \parallel \overline{DC}$
$\overline{AD} \parallel \overline{BC}$

10 in.

D 17 in. C

63.

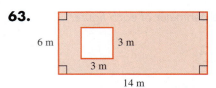

6 m   3 m

3 m

14 m

64.

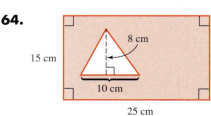

8 cm

15 cm

10 cm

25 cm

Solve each problem.

65. The area of a rectangle is 36 cm², and its length is 3 cm.
  a. Find its width.
  b. Find its perimeter.

66. The area of a parallelogram is 60 m², and its height is 15 m. Find the length of its base.

67. The area of a triangle is 54 ft², and the length of its base is 3 ft. Find the height.

68. The area of a square is 81 in².
  a. Find the length of a side.
  b. Find the perimeter.

69. The perimeter of a rectangle is 60 cm, and the width is 12 cm less than the length. Find the area of the rectangle.

70. The perimeter of a rectangle is 38 m, and the length is 1 m more than the width. Find the area of the rectangle.

71. The width of a rectangle is 6 in. less than its length, and its area is 16 in².
  a. Find its width and its length.
  b. Find its perimeter.

72. The length of a rectangle is 1 ft more than twice its width, and its area is 10 ft².
  a. Find its length and its width.
  b. Find its perimeter.

APPLICATIONS

73. FENCING A YARD   A man wants to enclose a rectangular yard with fencing that costs $12.50 a foot, including installation. Find the cost of enclosing the yard if its dimensions are 110 ft by 85 ft.

74. FRAMING A PICTURE   Find the cost of framing a rectangular picture with dimensions of 24 inches by 30 inches if framing material costs $8.46 per foot, including matting.

75. PLANTING A SCREEN   A woman wants to plant a pine-tree screen around three sides of her backyard. (See Illustration 7.) If she plants the trees 3 feet apart, how many trees will she need?

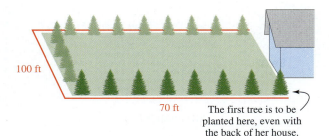

100 ft

70 ft

The first tree is to be planted here, even with the back of her house.

ILLUSTRATION 7

76. PLANTING MARIGOLDS   A gardener wants to plant a border of marigolds around the garden shown in Illustration 8, to keep out rabbits. How many plants will she need if she allows 6 inches between plants?

16 ft

20 ft

ILLUSTRATION 8

77. BUYING A FLOOR   Which is more expensive: A ceramic-tile floor costing $3.75 per square foot or linoleum costing $34.95 per square yard?

78. BUYING A FLOOR   Which is cheaper: A hardwood floor costing $5.95 per square foot or a carpeted floor costing $37.50 per square yard?

79. CARPETING A ROOM   A rectangular room is 24 feet long and 15 feet wide. At $30 per square yard, how much will it cost to carpet the room? (Assume no waste.)

80. CARPETING A ROOM   A rectangular living room measures 30 by 18 feet. At $32 per square yard, how much will it cost to carpet the room? (Assume no waste.)

81. TILING A FLOOR   A rectangular basement room measures 14 by 20 feet. Vinyl floor tiles that are 1 ft^2 cost $1.29 each. How much will the tile cost to cover the floor? (Disregard any waste.)

82. PAINTING A BARN   The north wall of a barn is a rectangle 23 feet high and 72 feet long. There are five windows in the wall, each 4 by 6 feet. If a gallon of paint will cover 300 ft^2, how many gallons of paint must the painter buy to paint the wall?

83. MAKING A SAIL   If nylon is $12 per square yard, how much would the fabric cost to make a triangular sail with a base of 12 feet and a height of 24 feet?

84. PAINTING A GABLE   The gable end of a warehouse is an isosceles triangle with a height of 4 yards and a base of 23 yards. It will require one coat of primer and one coat of finish to paint the triangle. Primer costs $17 per gallon, and the finish paint costs $23 per gallon. If one gallon covers 300 square feet, how much will it cost to paint the gable, excluding labor?

85. GEOGRAPHY   See Illustration 9. Use the dimensions of the trapezoid that is superimposed over the state of Nevada to estimate the area of the "Silver State."

ILLUSTRATION 9

86. COVERING A SWIMMING POOL   A swimming pool has the shape shown in Illustration 10. How many square meters of plastic sheeting will be needed to cover the pool? How much will the sheeting cost if it is $2.95 per square meter? (Assume no waste.)

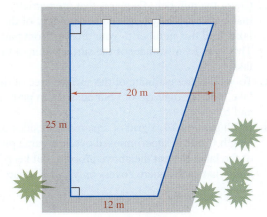

20 m

25 m

12 m

ILLUSTRATION 10

87. CARPENTRY   How many sheets of 4-foot-by-8-foot sheetrock are needed to drywall the inside walls on the first floor of the barn shown in Illustration 11? (Assume that the carpenters will cover each wall entirely and then cut out areas for the doors and windows.)

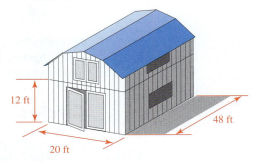

12 ft

20 ft

48 ft

ILLUSTRATION 11

88. CARPENTRY   If it costs $90 per square foot to build a one-story home in northern Wisconsin, estimate the cost of building the house with the floor plan shown in Illustration 12.

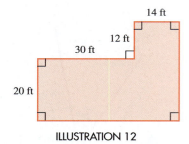

ILLUSTRATION 12

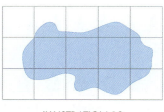

ILLUSTRATION 13

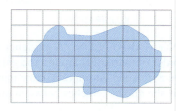

ILLUSTRATION 14

89. ESTIMATING SURFACE AREA   In Illustration 13, a grid is superimposed over a picture of a lake. Each square is 1 mile on a side and therefore represents 1 square mile.
  a. Count the number of squares that are completely within the boundary (shoreline) of the lake. This is an underestimate of the surface area of the lake.
  b. Count the number of squares that are partially inside and partially outside the boundary of the lake. Add this number to your answer from part a. This is an overestimate of the surface area of the lake.
  c. To get a better estimate of the surface area of the lake, find the *average* of your answers to parts a and b.
  d. In Illustration 14, a grid of squares with sides of length $\frac{1}{2}$ mile is superimposed over the same picture of the lake. Repeat the above process, but keep in mind that each square covers only $\frac{1}{4}$ of a square mile.

90. ESTIMATING AREA   See Illustration 15. Estimate the area of the sole plate of the iron by thinking of it as a combination of a trapezoid and a triangle.

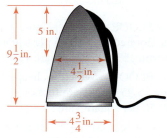

ILLUSTRATION 15

WRITING

91. Explain the difference between perimeter and area.

92. Why is it necessary that area be measured in square units?

---

7   *Circles*

In this section, you will learn about

- Circles • Circumference of a circle • Area of a circle • Arc length
- Area of a sector

INTRODUCTION. In this section, we will discuss the circle, one of the most useful geometric figures. In fact, the discoveries of fire and the circular wheel were two of the most important events in the history of the human race.

Circles

Circle   A **circle** is the set of all points in a plane that lie a fixed distance from a point called its **center.**

A segment drawn from the center of a circle to a point on the circle is called a **radius.** (The plural of *radius* is *radii.*) From the definition, it follows that all radii of the same circle are the same length.

A **chord** of a circle is a line segment that connects two points on the circle. A **diameter** is a chord that passes through the center of the circle. Since a diameter D of a circle is twice as long as a radius r, we have

$$D = 2r$$

Each of the previous definitions is illustrated in Figure 85, in which O is the center of the circle.

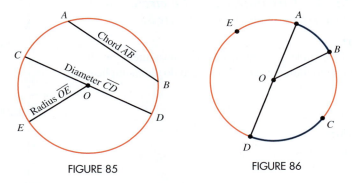

FIGURE 85   FIGURE 86

Any part of a circle is called an **arc.** In Figure 86, the part of the circle from point A to point B is $\overset{\frown}{AB}$, read as "arc AB." $\overset{\frown}{CD}$ is the part of the circle from point C to point D. An arc that is half of a circle is a **semicircle.**

Semicircle	A **semicircle** is an arc of a circle whose endpoints are the endpoints of a diameter.

If point O is the center of the circle in Figure 86, $\overline{AD}$ is a diameter and $\overset{\frown}{AED}$ is a semicircle. The middle letter E distinguishes semicircle $\overset{\frown}{AED}$ (the part of the circle from point A to point D that includes point E) from semicircle $\overset{\frown}{ABD}$ (the part of the circle from point A to point D that includes point B).

An arc that is shorter than a semicircle is a **minor arc.** An arc that is longer than a semicircle is a **major arc.** In Figure 86,

$\overset{\frown}{AE}$ is a minor arc   and $\overset{\frown}{ABE}$ is a major arc.

COMMENT It is often possible to name a major arc in more than one way. For example, in Figure 86, major arc $\overset{\frown}{ABE}$ is the part of the circle from point A to point E that includes point B. Two other names for the same major arc are $\overset{\frown}{ACE}$ and $\overset{\frown}{ADE}$.

Circumference of a circle

Since early history, mathematicians have known that the ratio of the distance around a circle (the circumference) divided by the length of its diameter is approximately 3. First Kings, Chapter 7 of the Bible describes a round bronze tank that was 15 feet from brim to brim and 45 feet in circumference, and $\frac{45}{15} = 3$. Today, we have a better value for this ratio, known as π (pi). If C is the circumference of a circle and D is the length of its diameter, then

$$\pi = \frac{C}{D} \qquad \text{where } \pi = 3.141592653589. \ . \ . \qquad \text{$\frac{22}{7}$ and 3.14 are often used as estimates of π.}$$

If we multiply both sides of $\pi = \frac{C}{D}$ by D, we have the following formula.

Circumference of a circle

The circumference of a circle is given by the formula

$$C = \pi D \quad \text{where } C \text{ is the circumference and } D \text{ is the length of the diameter}$$

Since a diameter of a circle is twice as long as a radius r, we can substitute $2r$ for D in the formula $C = \pi D$ to obtain another formula for the circumference C:

$$C = 2\pi r \quad 2\pi r \text{ means } 2 \cdot \pi \cdot r.$$

EXAMPLE 1   *Circumference of a circle.*   Find
the circumference of the circle shown in Figure 87.

Solution
The radius of the circle is 5 centimeters. We substitute 5 for r in the formula for circumference of a circle and do the multiplication.

$C = 2\pi\mathbf{r}$

$C = 2\pi(\mathbf{5})$

$C = 2(5)\pi$

$C = 10\pi$ Normally, when a product involves π, we rewrite it so that π is the last factor.

The circumference of the circle is exactly 10π cm. If we replace π with 3.14, we get an approximation of the circumference.

$C = 10\pi$

$C \approx 10(\mathbf{3.14})$

$C \approx 31.4$ To multiply by 10, move the decimal point in 3.14 one place to the right.

The circumference of the circle is approximately 31.4 cm.

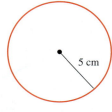

FIGURE 87

Self Check

Find the circumference of a circle that has a radius of 12 meters. Give the exact answer and an approximation, to the nearest tenth.

Answer: 24π m, 75.4 m   ■

Accent on Technology: **Calculating revolutions of a tire**

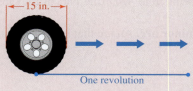

One revolution

FIGURE 88

When the $\boxed{\pi}$ key on a scientific calculator is pressed (on some models, the $\boxed{\text{2nd}}$ key must be pressed first), an approximation of π is displayed. To illustrate how to use this key, consider the following problem. How many times does the tire shown in Figure 88 revolve when a car makes a 25-mile trip?

We first find the circumference of the tire. From the figure, we see that the diameter of the tire is 15 inches. Since the circumference of a circle is the product of π and the length of its diameter, the tire's circumference is $\pi \cdot 15$ inches or 15π inches. (Normally, we rewrite a product such as $\pi \cdot 15$ so that π is the second factor.)

We then change the 25 miles to inches using two unit conversion factors.

$$\frac{25 \text{ miles}}{1} \cdot \frac{5{,}280 \text{ feet}}{1 \text{ mile}} \cdot \frac{12 \text{ inches}}{1 \text{ foot}} = 25 \cdot 5{,}280 \cdot 12 \text{ inches} \quad \text{The units of miles and feet divide out.}$$

The length of the trip is $25 \cdot 5{,}280 \cdot 12$ inches.

Finally, we divide the length of the trip by the circumference of the tire to get

$$\text{The number of revolutions of the tire} = \frac{25 \cdot 5{,}280 \cdot 12}{15\pi}$$

We can do the division using a scientific calculator.

Keystrokes $\boxed{(}$ 25 $\boxed{\times}$ 5280 $\boxed{\times}$ 12 $\boxed{)}$ $\boxed{\div}$ $\boxed{(}$ 15 $\boxed{\times}$ $\boxed{\pi}$ $\boxed{)}$ $\boxed{=}$

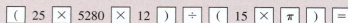

$$\boxed{33613.52398}$$

The tire makes about 33,614 revolutions.

EXAMPLE 2   *Finding the radius of a circle.*   The circumference of a circle is 50 inches. What is its radius?

Solution

We can find the radius of the circle by substituting 50 for C in the formula $C = 2\pi r$ and then solving for r.

$$C = 2\pi r$$

$$50 = 2\pi r$$

$$\frac{50}{2\pi} = \frac{2\pi r}{2\pi}$$ To undo the multiplication by 2π, divide both sides by 2π.

$$\frac{50}{2\pi} = r$$ This is the exact value of r.

$$r \approx 7.957747155$$ Use a calculator to do the division.

The radius of the circle is approximately 8 inches.

Self Check
Find the radius of a circle if it has a circumference of 25 inches.

Answer: about 4 in.   ■

EXAMPLE 3   *Architecture.*   A Norman window is constructed by adding a semicircular window to the top of a rectangular window. Find the perimeter of the Norman window shown in Figure 89.

Solution   The window is a combination of a rectangle and a semicircle. The perimeter of the rectangular part is

$$P_{\text{rectangular part}} = 8 + 6 + 8 = 22$$ Add only 3 sides.

The perimeter of the semicircle is one-half of the circumference of a circle that has a 6-foot diameter.

$$P_{\text{semicircle}} = \tfrac{1}{2}C$$

$$P_{\text{semicircle}} = \frac{1}{2}\pi D$$ Since we know the diameter, replace C with πD. We could also have replaced C with $2\pi r$.

$$= \frac{1}{2}\pi(6)$$ Substitute 6 for D.

$$\approx 9.424777961$$ Use a calculator.

The total perimeter is the sum of the two parts.

$$P_{\text{total}} \approx 22 + 9.424777961$$

$$\approx 31.424777961$$

To the nearest hundredth, the perimeter of the window is 31.42 feet.   ■

8 ft   8 ft

6 ft

FIGURE 89

Area of a circle

If we divide the circle shown in Figure 90(a) into an even number of pie-shaped pieces and then rearrange them as shown in Figure 90(b), we have a figure that looks like a parallelogram. The figure has a base b that is one-half the circumference of the circle, and its height h is about the same length as a radius of the circle.

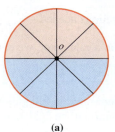

(a)

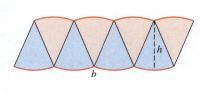

(b)

FIGURE 90

If we divide the circle into more and more pie-shaped pieces, the figure will look more and more like a parallelogram, and we can find its area by using the formula for the area of a parallelogram.

$$A = \textbf{\textit{bh}}$$

$$A = \frac{\textbf{1}}{\textbf{2}}\textbf{\textit{Cr}} \qquad \text{Substitute } \tfrac{1}{2} \text{ of the circumference for } b, \text{ and } r \text{ for the height.}$$

$$= \frac{1}{2}(2\pi r)r \quad \text{Make a substitution: } C = 2\pi r.$$

$$= \pi r^2 \qquad \text{Simplify: } \tfrac{1}{2} \cdot 2 = 1 \text{ and } r \cdot r = r^2.$$

Area of a circle	The **area of a circle** with radius r is given by the formula $$A = \pi r^2$$

EXAMPLE 4   *Area of a circle.*   To the nearest tenth, find the area of the circle in Figure 91.

Self Check

To the nearest tenth, find the area of a circle with a diameter of 12 feet.

Solution
Since the length of the diameter is 10 centimeters and the length of a diameter is twice the length of a radius, the length of the radius is 5 centimeters. To find the area of the circle, we substitute 5 for r in the formula for the area of a circle.

10 cm

FIGURE 91

$$A = \pi r^2$$

$$A = \pi(\textbf{5})^2 \quad \pi r^2 \text{ means } \pi \cdot r^2.$$

$$= \pi(25)$$

$$= 25\pi \quad \text{Write the product so that } \pi \text{ is the last factor.}$$

The exact area of the circle is 25π cm^2. We can use a calculator to approximate the area.

$$A \approx 78.53981634 \quad \text{Use a calculator to do the multiplication } 25 \cdot \pi.$$

Answer: 113.1 ft^2

To the nearest tenth, the area is 78.5 cm^2.

Accent on Technology: *Painting a helicopter landing pad*

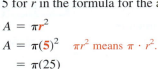

Orange paint is available in gallon containers at \$19 each, and each gallon will cover 375 ft^2. To calculate how much the paint will cost to cover a circular helicopter landing pad 60 feet in diameter, we first calculate the area of the helicopter pad.

$$A = \pi r^2$$

$$A = \pi(\textbf{30})^2 \quad \text{Substitute one-half of 60 for } r.$$

$$= 30^2\pi$$

The area of the pad is $30^2\pi$ ft^2. Since each gallon of paint will cover 375 ft^2, we can find the number of gallons of paint needed by dividing $30^2\pi$ by 375.

$$\text{Number of gallons needed} = \frac{30^2\pi}{375}$$

To do this work on a calculator, we enter these numbers and press these keys.

Keystrokes   30 $\boxed{x^2}$ $\boxed{\times}$ $\boxed{\pi}$ $\boxed{=}$ $\boxed{\div}$ 375 $\boxed{=}$ $\boxed{\text{7.539822369}}$

Because paint comes only in full gallons, the painter will need to purchase 8 gallons. The cost of the paint will be 8(\$19), or \$152.

EXAMPLE 5   *Crop circles.*   Geometric shapes like that in Figure 92 have been appearing in the fields of England since the mid-1970s. Since then, *crop circles,* as they are called, have appeared in over 20 countries. If one crop circle was reported to cover an area of 70,000 ft^2, what was its diameter?

FIGURE 92

Solution

We can first find the radius of the crop circle by substituting 70,000 for A in the formula $A = \pi r^2$ and then solving for r.

$$A = \pi r^2$$

$$70{,}000 = \pi r^2$$

$$\frac{70{,}000}{\pi} = \frac{\pi r^2}{\pi} \qquad \text{To undo the multiplication by } \pi, \text{ divide both sides by } \pi.$$

$$\frac{70{,}000}{\pi} = r^2$$

To find r, we must find a number that, when squared, is $\frac{70{,}000}{\pi}$. There are two such numbers, one positive and one negative; they are the square roots of $\frac{70{,}000}{\pi}$. Since r is the radius of a circle, r cannot be negative. For this reason, we only find the positive square root of $\frac{70{,}000}{\pi}$ to determine r.

$$\sqrt{\frac{70{,}000}{\pi}} = r \qquad\qquad \text{This is the exact value of } r.$$

$$r \approx 149.270533 \quad \text{Use a calculator.}$$

To find the diameter of the crop circle, we multiply the radius by 2.

$$D = 2r \approx 2(149.270533) \approx 298.5$$

The diameter of the crop circle was approximately 300 ft.   ■

EXAMPLE 6   *Finding the area.*   Find the shaded area in Figure 93.

Solution   The figure is a combination of a triangle and two semicircles.

The area of the triangle is

$$A_{\text{right triangle}} = \frac{1}{2}bh = \frac{1}{2}(6)(8) = \frac{1}{2}(48) = 24$$

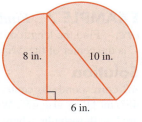

8 in.   10 in.

6 in.

FIGURE 93

The area enclosed by the smaller semicircle is

$$A_{\text{smaller semicircle}} = \frac{1}{2}\pi r^2 = \frac{1}{2}\pi(4)^2 = \frac{1}{2}\pi(16) = 8\pi$$

The area enclosed by the larger semicircle is

$$A_{\text{larger semicircle}} = \frac{1}{2}\pi r^2 = \frac{1}{2}\pi(5)^2 = \frac{1}{2}\pi(25) = 12.5\pi$$

The total area is

$$A_{\text{total}} = 24 + 8\pi + 12.5\pi \approx 88.4026494 \quad \text{Use a calculator.}$$

To the nearest hundredth, the area is 88.40 in.^{2} ■

Arc length

A **central angle** of a circle is an angle whose vertex is the center of the circle. In Figure 94, the circle has center C. In this circle, $\angle ACB$, with vertex C, is a central angle.

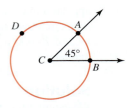

FIGURE 94

In Figure 94, the sides of central angle $\angle ACB$ intersect the circle to create minor arc $\overset{\frown}{AB}$ and major arc $\overset{\frown}{ADB}$. The **degree measure of an arc** is defined to be equal to the degree measure of its corresponding central angle. Since $m(\angle ACB) = 45°$, the measure of $\overset{\frown}{AB}$ is $45°$, and we can write $m(\overset{\frown}{AB}) = 45°$.

Recall that an angle of one revolution measures $360°$. We can use this fact to find the measure of major arc $\overset{\frown}{ADB}$ in Figure 94. In general, the measure of a major arc is simply the difference of $360°$ and the measure of its associated minor arc.

$$m(\overset{\frown}{ADB}) = 360° - \mathbf{m(\overset{\frown}{AB})} = 360° - \mathbf{45°} = 315°$$

We can use the following formula to find the length of an arc of a circle.

Length of an arc

> If an arc has measure q (in degrees) and radius r, its length L is given by
>
> $$L = \frac{q}{360°} \cdot 2\pi r \quad \text{\textcolor{red}{L and r have the same units.}}$$

 COMMENT   Note that the length of an arc of a circle is a fraction of the circumference of the circle ($2\pi r$). The fraction is the ratio of the measure of the arc to one complete revolution, $360°$.

EXAMPLE 7 *Finding the length of a circular arc.*   Find the length of minor arc $\overset{\frown}{AB}$ shown in Figure 95. C is the center of the circle.

Solution
$\overset{\frown}{AB}$ corresponds to central angle $\angle ACB$. Since $m(\angle ACB) = 45°$, $m(\overset{\frown}{AB})$ is also $45°$. We can find the length of $\overset{\frown}{AB}$ using the arc length formula, where q is $45°$ and the radius r of the circle is 18 inches.

$$L = \frac{\mathbf{q}}{360°} \cdot 2\pi \mathbf{r}$$

$$L = \frac{\mathbf{45°}}{360°} \cdot 2\pi(\mathbf{18}) \quad \text{\textcolor{red}{Substitute 45° for q and 18 for r.}}$$

$$= \frac{\overset{1}{\cancel{45°}}}{8 \cdot \underset{1}{\cancel{45°}}} \cdot 36\pi \quad \text{\textcolor{red}{Factor 360° as 8 · 45° and divide out the common factor of 45°. Write $2\pi(18)$ so that π is the last factor: 36π.}}$$

$$= \frac{1}{8} \cdot 36\pi \quad \text{\textcolor{red}{Note that the length of the arc is $\frac{1}{8}$ of the circumference of the circle, 36π.}}$$

FIGURE 95

Self Check
Find the length of minor arc $\overset{\frown}{MN}$ shown in the figure below. C is the center of the circle.

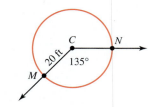

$$= \frac{36\pi}{8} \qquad \text{Multiply: } \frac{1}{8} \cdot 36\pi = \frac{1}{8} \cdot \frac{36\pi}{1}.$$

$$= \frac{\overset{1}{\cancel{4}} \cdot 9\pi}{\underset{1}{\cancel{4}} \cdot 2} \qquad \text{Factor } 36\pi \text{ as } 4 \cdot 9\pi \text{ and } 8 \text{ as } 4 \cdot 2. \text{ Divide out the common factor of } 4 \text{ in the numerator and denominator.}$$

$$= \frac{9\pi}{2}$$

The length of $\overset{\frown}{AB}$ is exactly $\frac{9\pi}{2}$ inches. To the nearest tenth, $\frac{9\pi}{2}$ inches ≈ 14.1 inches.　　|　**Answer:**　15π ft ≈ 47.1 ft　■

Area of a sector

The shaded region in Figure 96 is called a **sector.** The sector in the figure has a radius of 20 meters, and the measure of its associated arc is 60°.

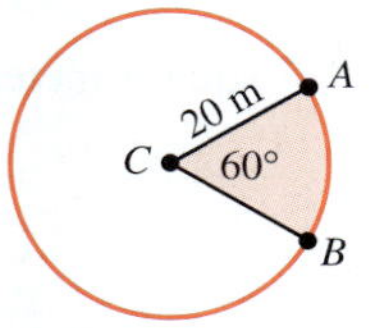

FIGURE 96

We can find the area of the sector using the following formula.

Area of a sector

If a sector has radius r, and its associated arc has measure q, its area is given by

$$A = \frac{q}{360°} \cdot \pi r^2$$

 COMMENT   Note that the area of a sector is a fraction of the area of the circle (πr^2). The fraction is the ratio of the associated arc measure to one complete revolution, 360°.

EXAMPLE 8　*Finding the area of a sector.*　Find the area of sector shown in Figure 97. C is the center of the circle.

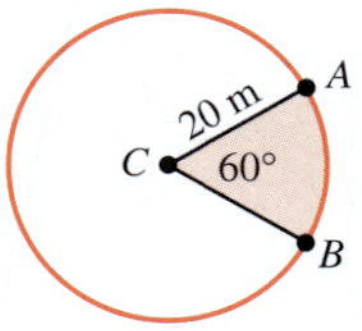

FIGURE 97

Solution

$\overset{\frown}{AB}$ corresponds to central angle $\angle ACB$. Since m($\angle ACB$) = 60°, m($\overset{\frown}{AB}$) is also 60°. We can find the area of the sector using the area formula, where q is 60° and r (the radius of the circle) is 20 meters.

$$A = \frac{q}{360°} \cdot \pi r^2$$

$$A = \frac{60°}{360°} \cdot \pi (20)^2 \qquad \text{Substitute 60° for } q \text{ and 20 for } r.$$

Self Check

Find the area of the sector shown below.

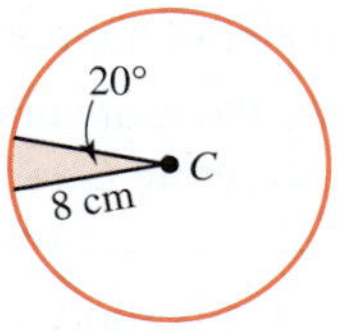

$$A = \frac{60°}{360°} \cdot \pi(400)$$ Evaluate the exponential expression: $(20)^2 = 400$.

$$A = \frac{\overset{1}{\cancel{60°}}}{6 \cdot \underset{1}{\cancel{60°}}} \cdot 400\pi$$ Factor 360° as 6 · 60° and divide out the common factor of 60°. Write $\pi(400)$ so that π is the last factor: 400π.

$$= \frac{1}{6} \cdot 400\pi$$ Note that the area of the sector is $\frac{1}{6}$ of the area of the circle, 400π.

$$= \frac{400\pi}{6}$$ Multiply: $\frac{1}{6} \cdot 400\pi = \frac{1}{6} \cdot \frac{400\pi}{1}$.

$$= \frac{\overset{1}{\cancel{2}} \cdot 200\pi}{\underset{1}{\cancel{2}} \cdot 3}$$ Factor 400π as $2 \cdot 200\pi$ and 6 as $2 \cdot 3$. Divide out the common factor of 2 in the numerator and denominator.

$$= \frac{200\pi}{3}$$

The area of the sector is exactly $\frac{200\pi}{3}$ square meters. To the nearest tenth, $\frac{200\pi}{3}$ m² ≈ 209.4 m².

Answer: $\frac{32\pi}{9}$ cm² ≈ 11.2 cm²

■

STUDY SET   Section 7

VOCABULARY   *Fill in the blanks.*

1. A segment drawn from the center of a circle to a point on the circle is called a _____.

2. A segment joining two points on a circle is called a _____.

3. A _____ is a chord that passes through the center of a circle.

4. An arc that is one-half of a complete circle is a _____.

5. The distance around a circle is called its _____.

6. The surface enclosed by a circle is called its _____.

7. A diameter of a circle is _____ as long as a radius.

8. Suppose the *exact* circumference of a circle is 3π feet. When we write $C \approx 9.42$ feet, we are giving an _____ of the circumference.

9. An arc that is shorter than a semicircle is called a _____ arc. An arc that is longer than a semicircle is called a _____ arc.

10. A _____ angle of a circle is an angle whose vertex is the center of the circle.

11. The degree measure of an _____ is defined to be the degree measure of its corresponding central angle.

12. The shaded region in Illustration 1 is called a _____.

ILLUSTRATION 1

CONCEPTS   *In Exercises 13–20, refer to Illustration 2, where O is the center of the circle.*

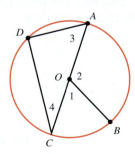

ILLUSTRATION 2

13. Name each radius.

14. Name a diameter.

15. Name each chord.

16. Name each minor arc.

17. Name each semicircle.

18. Name major arc $\overset{\frown}{ABD}$ in another way.

19. The sides of what central angle intersect the circle to create minor arc $\overset{\frown}{BC}$?

20. The sides of what central angle intersect the circle to create major arc $\overset{\frown}{ADB}$?

21. a. If you know the radius of a circle, how can you find its diameter?
    b. If you know the diameter of a circle, how can you find its radius?

22. One complete revolution is how many degrees?

23. Suppose the two "legs" of the compass shown in Illustration 3 are adjusted so that the distance between the pointed ends is 1 inch. Then a circle is drawn.
    a. What will the radius of the circle be?
    b. What will the diameter of the circle be?
    c. What will the circumference of the circle be? Give an exact answer and an approximation.
    d. What will the area of the circle be? Give an exact answer and an approximation.

ILLUSTRATION 3

24. Suppose we find the distance around a can and the distance across the can using a measuring tape, as shown in Illustration 4. Then we make a comparison, in the form of a ratio:

$$\frac{\text{The distance around the can}}{\text{The distance across the top of the can}}$$

After we do the indicated division, the result will be close to what number?

ILLUSTRATION 4

25. When evaluating $\pi(6)^2$, what operation should be performed first?

26. Round $\pi = 3.141592653589.\ .\ .$ to the nearest hundredth.

Refer to Illustration 5. X is the center of the circle.

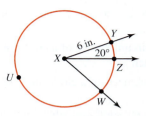

ILLUSTRATION 5

27. What is the diameter of the circle?

28. What is m($\overset{\frown}{YZ}$)? What is m($\overset{\frown}{YUZ}$)?

29. Name $\overset{\frown}{YUZ}$ in another way.

30. What is m($\overset{\frown}{YWZ}$)?

31. If m($\overset{\frown}{ZW}$) = 42°, what is m($\angle \overset{\frown}{ZYW}$)?

32. If m($\overset{\frown}{ZW}$) = 42°, what is m($\overset{\frown}{ZUW}$)?

33. What is the central angle associated with minor arc $\overset{\frown}{YW}$?

34. What is the radius of the sector associated with central angle $\angle ZXW$?

35. a. On the given circle, draw central angle $\angle ABC$ and label it completely.

    b. What is the formula that gives the length of $\overset{\frown}{AC}$?
    c. What part of the formula represents the entire circumference of the circle?
    d. What fraction of the entire circumference does this formula find?

36. a. On the given circle, draw and shade the sector associated with central angle $\angle ABC$. Label the figure completely.

    b. What is the formula that gives the area of the sector?
    c. What part of the formula represents the entire area of the circle?
    d. What fraction of the entire area does this formula find?

NOTATION *Fill in the blanks.*

37. The symbol $\overset{\frown}{AB}$ is read as _____.

38. To the nearest hundredth, the value of π is _____.

39. The formula for the circumference of a circle is
_____ or _____.

40. The formula $A = \pi r^2$ gives the area of a
_____.

41. If C is the circumference of a circle and D is its
diameter, then $\frac{C}{D} =$ ___.

42. If D is the diameter of a circle and r is its radius, then
$D =$ ___ r.

43. a. In the expression $2\pi r$, what operations are
indicated?
b. In the expression πr^2, what operations are
indicated?

44. Write each expression in better form.

a. $\pi(8)$ **b.** $2\pi(7)$ **c.** $\pi \cdot \dfrac{25}{3}$

45. Simplify each fraction.

a. $\dfrac{90°}{360°}$ **b.** $\dfrac{4\pi}{8}$ **c.** $\dfrac{27\pi}{30}$

46. a. What does $m(\overset{\frown}{AB})$ mean?
b. What does $m(\overset{\frown}{DEF})$ mean?

PRACTICE

47. Find the radius of a circle that has a circumference of
16π inches.

48. Find the radius of a circle that has a circumference of
30π meters.

49. Find the diameter of a circle that has a circumference of
5π centimeters.

50. Find the radius of a circle that has a circumference of
9π yards.

*In Exercises 51–58, solve each problem. Round your
answer to the nearest tenth.*

51. Find the circumference of a circle that has a diameter of
12 inches.

52. Find the circumference of a circle that has a radius of
20 feet.

53. Find the diameter of a circle that has a circumference of
113 meters.

54. Find the radius of a circle that has a circumference of
157 meters.

55. Find the circumference of the circle in Illustration 6.

ILLUSTRATION 6

56. Find the circumference of the semicircle in
Illustration 7.

25 cm

ILLUSTRATION 7

57. Find the circumference of the circle in Illustration 8 if
the square has sides of length 6 inches.

ILLUSTRATION 8

58. Find the circumference of the semicircle in Illustration
9 if the length of the rectangle is 8 feet.

ILLUSTRATION 9

Find the perimeter of each figure to the nearest hundredth.

59.
8 ft
3 ft

60.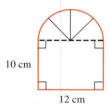
10 cm
12 cm

61.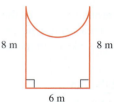
8 m   8 m
6 m

62.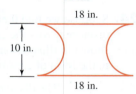
18 in.
10 in.
18 in.

*In Exercises 63–78, solve each problem. If an answer is
not exact, round it to the nearest tenth.*

63. Find the radius of a circle that has an area of 49π ft^2.

64. Find the radius of a circle that has an area of 64π cm^2.

65. Find the diameter of a circle that has an area of
$\dfrac{25\pi}{16}$ yd^2.

66. Find the diameter of a circle that has an area of
$\dfrac{36\pi}{25}\pi$ mi^2.

67. Find the area of a circle with radius 15 feet.

68. Find the area of a circle with radius 3.2 inches.

69. Find the area of a circle with diameter 50 inches.

70. Find the area of a circle with diameter 1 meter.

71. The area of a circle is 28 ft². What is its radius?

72. The area of a circle is 9.9 yd². What is its radius?

73. The area of a circle is 4.4 m².
   a. What is its radius?
   b. What is its diameter?
   c. What is its circumference?

74. The area of a circle is 150 cm².
   a. What is its radius?
   b. What is its diameter?
   c. What is its circumference?

Find the area of each circle.

75.

3 in.

76.
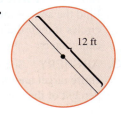
12 ft

77. Find the area of the circle in Illustration 10 if the square has sides of length 9 millimeters.

ILLUSTRATION 10

78. Find the area of the shaded semicircular region in Illustration 11.

6.5 mi

ILLUSTRATION 11

Find the total area of each figure to the nearest tenth.

79.
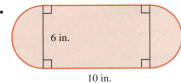
6 in.

10 in.

80.
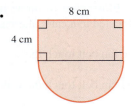
8 cm

4 cm

81.
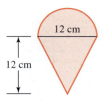
12 cm

12 cm

82.
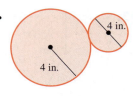
4 in.

4 in.

Find the area of each shaded region to the nearest tenth.

83.
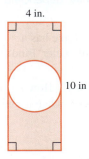
4 in.

10 in

84.
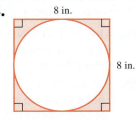
8 in.

8 in.

85.
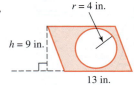
r = 4 in.

h = 9 in.

13 in.

86.
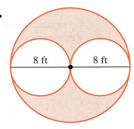
8 ft   8 ft

Find the exact length of minor arc $\overarc{AB}$. Then approximate it to the nearest tenth.

87.
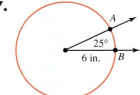
A
25°
6 in.   B

88.
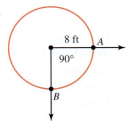
8 ft   A
90°
B

89.
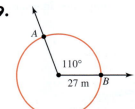
A
110°
27 m   B

90.
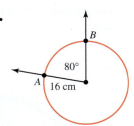
B
80°
A   16 cm

Find the exact area of the shaded sector. Then approximate it to the nearest tenth.

91.

10 ft
30°

92.

45°
9 in.

93.

94.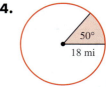

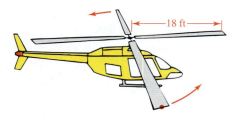

APPLICATIONS   *Give each answer to the nearest hundredth. Answers may vary slightly, depending on which approximation of π is used.*

95. AREA OF ROUND LAKE   Round Lake has a circular shoreline that is 2 miles in diameter. Find the area of the lake.

96. HELICOPTER   Refer to Illustration 12. How far does a point on the tip of a rotor blade travel when it makes one complete revolution?

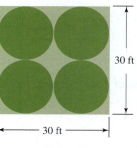

ILLUSTRATION 12

97. GIANT SEQUOIA   The largest sequoia tree is the General Sherman Tree in Sequoia National Park in California. In fact, it is considered to be the largest living thing in the world. According to the *Guinness Book of World Records,* it has a circumference of 102.6 feet, measured $4\frac{1}{2}$ feet above the ground. What is the diameter of the tree at that height?

98. TRAMPOLINE   See Illustration 13. The distance from the center of the trampoline to the edge of its steel frame is 7 feet. The protective padding covering the springs is 15 inches wide. Find the area of the circular jumping surface of the trampoline, in square feet.

ILLUSTRATION 13

99. JOGGING   Joan wants to jog 10 miles on a circular track $\frac{1}{4}$ mile in diameter. How many times must she circle the track?

100. FIXING THE ROTUNDA   The rotunda at a state capitol is a circular area 100 feet in diameter. The legislature wishes to appropriate money to have the floor of the rotunda tiled. The lowest bid is $83 per square yard, including installation. How much must the legislature spend?

101. BANDING THE EARTH   A steel band is drawn tightly about the Earth's equator. The band is then loosened by increasing its length by 10 feet, and the resulting slack is distributed evenly along the band's entire length. How far above the Earth's surface is the band? (*Hint:* You don't need to know the Earth's circumference.)

102. CONCENTRIC CIRCLES   Two coplanar circles are called **concentric circles** if they have the same center. Find the area of the band between two concentric circles if their diameters are 10 centimeters and 6 centimeters.

103. ARCHERY   See Illustration 14. Find the area of the entire target and the area of the bull's eye. What percent of the area of the target is the bull's eye?

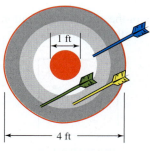

ILLUSTRATION 14

104. LANDSCAPE DESIGN   See Illustration 15. How many square feet of lawn does not get watered by the sprinklers at the center of each circle?

ILLUSTRATION 15

105. AUTOMOTIVE REPAIR   Illustration 16 shows how a fan belt turns pulleys connected to a car's alternator and water pump.
   a. How many inches of the fan belt touch the alternator pulley?
   b. How many inches of the fan belt touch the water pump pulley?

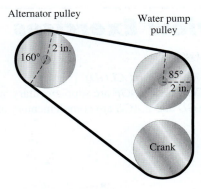

Alternator pulley   Water pump pulley

160° 2 in.

85°
2 in.

Crank

ILLUSTRATION 16

106. CLOCKS   Illustration 17 shows the minute hand of a clock moving from 12 to 3.
   a. What angle does the minute hand sweep out?
   b. If the minute hand is 5 inches long, how much area does it sweep out?

ILLUSTRATION 17

WRITING

107. Explain what is meant by the circumference of a circle.

108. Explain what is meant by the area of a circle.

109. Explain the meaning of π.

110. Distinguish between a major arc and a minor arc.

111. Explain what it means for a car to have a small turning radius.

112. The word *circumference* means the distance around a circle. In your own words, explain what is meant by each of the following sentences.
   a. A boat owner's dream was to *circumnavigate* the globe.
   b. The teenager's parents felt that he was always trying to *circumvent* the rules.
   c. The class was shown a picture of a circle *circumscribed* about an equilateral triangle.

Sections 1-7 Cumulative Review Exercises

1. What are the three undefined words in geometry?

2. Draw each geometric figure and label it.
  a. $\overleftrightarrow{AS}$
  b. $\overline{GH}$
  c. $\overrightarrow{MN}$
  d. $\angle BCA$

3. Are $\overrightarrow{AB}$ and $\overrightarrow{BA}$ the same ray?

4. Give four ways to name the angle shown in Illustration 1.

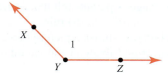

ILLUSTRATION 1

5. Measure each angle with a protractor. Then tell whether it is an acute, right, obtuse, or straight angle.
  a.
  b.

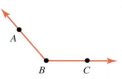

  c.
  d.

6. Fill in the blanks.
a. If $\angle ABC \cong \angle DEF$, then the angles have the same _____ .

b. Two congruent segments have the same _____ .

c. Two different points determine one _____ .
d. Two angles are _____ if the sum of their measures is 90°.

7. Refer to Illustration 2. What is the midpoint of $\overline{BE}$?

ILLUSTRATION 2

8. Refer to Illustration 3 and tell whether each statement is true or false.
  a. $\angle AGF$ and $\angle BGC$ are vertical angles.
  b. $\angle EGF$ and $\angle DGE$ are adjacent angles.

c. $m(\angle AGB) = m(\angle EGD)$.
d. $\angle CGD$ and $\angle DGF$ are supplementary angles.
e. $\angle EGD$ and $\angle AGB$ are complementary angles.

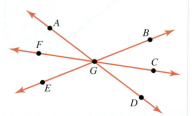

ILLUSTRATION 3

9. Find x. Then find $m(\angle ABD)$ and $m(\angle CBE)$.

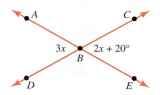

10. The measure of an angle is 20° more than three times its supplement's measure. What is the measure of the angle?

11. Refer to Illustration 4. Fill in the blanks.
  a. l_1 intersects two coplanar lines. It is called a _____ .

  b. $\angle 4$ and _____ are alternate interior angles.
  c. $\angle 3$ and _____ are corresponding angles.

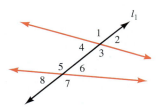

ILLUSTRATION 4

12. In Illustration 5, $l_1 \parallel l_2$ and $m(\angle 2) = 25°$. Find the measures of the other angles.

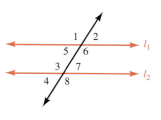

ILLUSTRATION 5

13. Find x. Then determine the measure of each angle that is labeled in the figure.

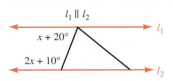

14. Are l_1 and l_2 parallel? If so, explain why.

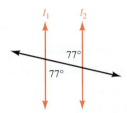

15. For each polygon, give the number of sides it has, tell its name, and then give the number of vertices it has.

a.

b.

c.

d.

16. Classify each triangle as an equilateral triangle, an isosceles triangle, or a scalene triangle.

a.

b.

c.

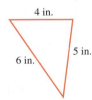

d.

17. Find x.

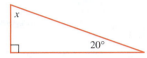

18. The measure of each base angle of an isosceles triangle is 30° more than the measure of the vertex angle. Find the measure of each angle.

19. What three quadrilaterals are used in the design of the Chevrolet ad in Illustration 6?

ILLUSTRATION 6

20. Draw each figure.
   a. Rhombus $ABCD$ **b.** Trapezoid $QRST$

21. Refer to rectangle $EFGH$ in Illustration 7.
   a. Find m($\overline{HG}$).   **b.** Find m($\overline{FH}$).
   c. Find m($\angle FGH$).   **d.** Find m($\overline{EH}$).

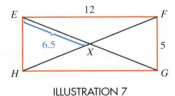

ILLUSTRATION 7

22. Refer to isosceles trapezoid $QRST$ in Illustration 8.
   a. Find m($\overline{RS}$).   **b.** Find x.
   c. Find y.   **d.** Find z.

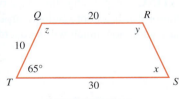

ILLUSTRATION 8

23. Find the sum of the measures of the angles of a decagon.

24. a. Find the number of sides of a regular polygon if one of its angles has a measure of 120°.

   b. What is the measure of an exterior angle of the polygon in part a?

25. Find the perimeter of the figure.

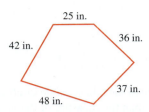

26. The perimeter of an equilateral triangle is 45.6 m. Find the length of each side.

27. The perimeter of a rectangle is 34 ft. The length is 3 feet less than four times the width. Find the length and width.

28. Find the area of the shaded part of the figure.

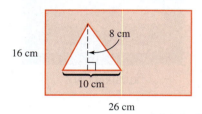

29. Find the area of the parallelogram in Illustration 9.

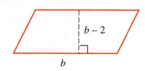

ILLUSTRATION 9

30. Draw an altitude to the base of the triangle.

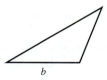

31. DECORATING A patio has the shape of a trapezoid. See Illustration 10. If indoor/outdoor carpeting sells for $18 a square yard, how much will it cost to carpet the patio?

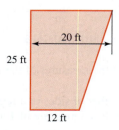

ILLUSTRATION 10

32. How many square inches are in one square foot?

33. A student said the area of the square in Illustration 11 was 25^2 ft. Is this correct? Explain why or why not.

ILLUSTRATION 11

34. Find the area of the rectangle in Illustration 12.

ILLUSTRATION 12

35. Refer to Illustration 13, where O is the center of the circle.

a. Name each chord.

b. Name each diameter.

c. Name each radius.

d. The sides of central angle $\angle DOB$ intersect the circle to form what minor and what major arc?

e. What type of figure is $\overset{\frown}{ACB}$?

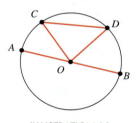

ILLUSTRATION 13

In Exercises 36–41, find each answer to the nearest tenth.

36. Find the circumference of a circle with a diameter of 21 centimeters.

37. Find the perimeter of the figure shown in Illustration 14.

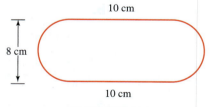

ILLUSTRATION 14

38. TOYS The circumference of a Hula Hoop is 106.8 inches. What is its diameter?

39. HISTORY Stonehenge is a prehistoric monument in England, believed to have been built by the Druids. The site, 30 meters in diameter, consists of a circular arrangement of stones, as shown in Illustration 15. What area does the monument cover?

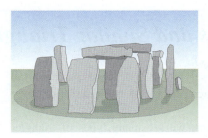

ILLUSTRATION 15

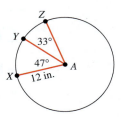

ILLUSTRATION 17

40. Find the area of the shaded part of the figure shown in Illustration 16 if the square has sides of length 4.5 miles.

ILLUSTRATION 16

43. Find the exact length of $\widehat{YZ}$ in Illustration 17. Then approximate it to the nearest tenth.

44. SNOWMAKING   A snowmaking machine sends out a spray that covers an area shaped like a sector of a circle. See Illustration 18. Find the area of the sector.

41. HOT TUB   The surface area of the water in a circular hot tub is 38.5 ft².

   a. What is its radius?

   b. What is its diameter?

   c. What is its circumference?

42. Refer to Illustration 17, where *A* is the center of the circle.

   a. What is m($\widehat{XY}$)?

   b. What is m($\widehat{YZX}$)?

   c. What is m($\widehat{YZ}$)?

ILLUSTRATION 18

8 *Congruent Triangles and Similar Triangles*

In this section, you will learn about

* Congruent triangles • Congruence properties • Similar triangles

INTRODUCTION. In our everyday lives, we see many types of triangles. Triangular-shaped kites, sails, roofs, tortilla chips, and ramps are just a few examples. In this section, we will discuss how to formally compare the size and shape of two given triangles. From this comparison, we can make deductions about their respective side lengths and angle measures.

Congruent triangles

Simply put, two geometric figures are **congruent** if they have the same shape and size. For example, if $\triangle ABC$ and $\triangle DEF$ in Figure 98 are congruent, we can write

$\triangle ABC \cong \triangle DEF$ Read as "Triangle *ABC* is congruent to triangle *DEF*."

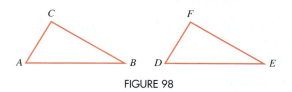

FIGURE 98

One way to determine whether two triangles are congruent is to see if one triangle can be moved onto the other triangle in such a way that it fits exactly. When we write $\triangle ABC \cong \triangle DEF$, we are showing how the vertices of one triangle are matched to the vertices of the other triangle to obtain a "perfect fit." We call this matching of points a **correspondence.**

$$\triangle ABC \cong \triangle DEF$$

$A \leftrightarrow D$ Read as "Point *A* corresponds to point *D*."
$B \leftrightarrow E$ Read as "Point *B* corresponds to point *E*."
$C \leftrightarrow F$ Read as "Point *C* corresponds to point *F*."

When we establish a correspondence between the vertices of two congruent triangles, we also establish a correspondence between the angles and the sides of the triangles. Corresponding angles and corresponding sides of congruent triangles are called **corresponding parts.** Corresponding parts of congruent triangles are always congruent. That is, corresponding parts of congruent triangles always have the same measure. For the congruent triangles in Figure 98, we have

$m(\angle A) = m(\angle D)$ $m(\angle B) = m(\angle E)$ $m(\angle C) = m(\angle F)$
$m(\overline{BC}) = m(\overline{EF})$ $m(\overline{AC}) = m(\overline{DF})$ $m(\overline{AB}) = m(\overline{DE})$

Congruent triangles   Two triangles are congruent if and only if their vertices can be matched so that the corresponding sides and the corresponding angles are congruent.

EXAMPLE 1   *Corresponding parts of congruent triangles.*
Refer to Figure 99, where $\triangle XYZ \cong \triangle PQR$.

a. Name the six congruent corresponding parts of the triangles.

b. Find $m(\angle P)$.

c. Find $m(\overline{XZ})$.

Solution

a. The correspondence between vertices is

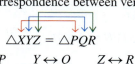

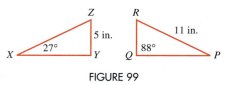

FIGURE 99

$$\triangle XYZ = \triangle PQR$$

$$X \leftrightarrow P \qquad Y \leftrightarrow Q \qquad Z \leftrightarrow R$$

Corresponding parts of congruent triangles are congruent. Therefore, the congruent corresponding angles are

$$\angle X \cong \angle P, \qquad \angle Y \cong \angle Q, \qquad \angle Z \cong \angle R$$

The congruent corresponding sides are

$$\overline{YZ} \cong \overline{QR}, \qquad \overline{XZ} \cong \overline{PR}, \qquad \overline{XY} \cong \overline{PQ}$$

b. From the figure, we see that m($\angle X$) = 27°. Since $\angle X \cong \angle P$, it follows that m($\angle X$) = 27°.

c. From the figure, we see that m($\overline{PR}$) = 11 inches. Since $\overline{XZ} \cong \overline{PR}$, it follows that m($\overline{XZ}$) = 11 inches.   ■

Congruence properties

Sometimes it is possible to conclude that two triangles are congruent without having to show that three pairs of corresponding angles are congruent and three pairs of corresponding sides are congruent. To do so, we apply one of the following properties.

SSS property

> If three sides of one triangle are congruent to three sides of a second triangle, the triangles are congruent.

We can show that the triangles in Figure 100 are congruent by the SSS property:

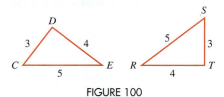

FIGURE 100

$\overline{CD} \cong \overline{ST}$	Since m($\overline{CD}$) = 3 and m($\overline{ST}$) = 3, the segments are congruent.
$\overline{DE} \cong \overline{TR}$	Since m($\overline{DE}$) = 4 and m($\overline{TR}$) = 4, the segments are congruent.
$\overline{EC} \cong \overline{RS}$	Since m($\overline{EC}$) = 5 and m($\overline{RS}$) = 5, the segments are congruent.

Therefore, $\triangle CDE \cong \triangle STR$.

SAS property

> If two sides and the angle between them in one triangle are congruent, respectively, to two sides and the angle between them in a second triangle, the triangles are congruent.

We can show that the triangles in Figure 101 are congruent by the SAS property:

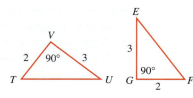

FIGURE 101

$\overline{TV} \cong \overline{FG}$ Since $m(\overline{TV}) = 2$ and $m(\overline{FG}) = 2$, the segments are congruent.

$\angle V \cong \angle G$ Since $m(\angle V) = 90°$ and $m(\angle G) = 90°$, the angles are congruent.

$\overline{UV} \cong \overline{EG}$ Since $m(\overline{UV}) = 3$ and $m(\overline{EG}) = 3$, the segments are congruent.

Therefore, $\triangle TVU \cong \triangle FGE$.

ASA property

> If two angles and the side between them in one triangle are congruent, respectively, to two angles and the side between them in a second triangle, the triangles are congruent.

We can show that the triangles in Figure 102 are congruent by the ASA property:

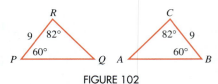

FIGURE 102

$\angle P \cong \angle B$ Since $m(\angle P) = 60°$ and $m(\angle B) = 60°$, the angles are congruent.

$\overline{PR} \cong \overline{BC}$ Since $m(\overline{PR}) = 9$ and $m(\overline{BC}) = 9$, the segments are congruent.

$\angle R \cong \angle C$ Since $m(\angle R) = 82°$ and $m(\angle C) = 82°$, the angles are congruent.

Therefore, $\triangle PQR \cong \triangle BAC$.

 COMMENT   There is no SSA property. To illustrate this, consider the triangles in Figure 103. Two sides and an angle of $\triangle ABC$ are congruent to two sides and an angle of $\triangle DEF$. But the congruent angle is not between the congruent sides.

We refer to this situation as SSA. Obviously, the triangles are not congruent, because they are not the same shape and size.

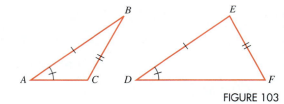

FIGURE 103

The tick marks indicate congruent parts. That is, the sides with one tick mark are the same length, the sides with two tick marks are the same length, and the angles with one tick mark have the same measure.

EXAMPLE 2   *Determining whether triangles are congruent.*
Explain why the triangles in Figure 104 are congruent.

Solution   Since vertical angles are congruent,

$$\angle 1 \cong \angle 2$$

From the figure, we see that

$$\overline{AC} \cong \overline{EC} \quad \text{and} \quad \overline{BC} \cong \overline{DC}$$

Since two sides and the angle between them in one triangle are congruent, respectively, to two sides and the angle between them in a second triangle, $\triangle ABC \cong \triangle EDC$ by the SAS property.

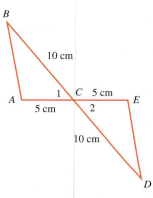

FIGURE 104

EXAMPLE 3   *Triangles having a common side.*

Are △*RST* and △*RUT* in Figure 105 congruent?

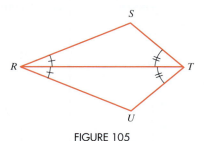

FIGURE 105

Self Check

Are the triangles in the following figure congruent?

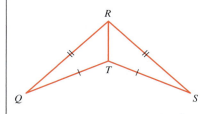

Solution

From the markings on the figure, we know that two pairs of angles are congruent.

∠*SRT* ≅ ∠*URT*   These angles are marked with 1 tick mark, which indicates that they have the same measure.

∠*STR* ≅ ∠*UTR*   These angles are marked with 2 tick marks, which indicates that they have the same measure.

From the figure, we see that the triangles have side $\overline{RT}$ in common. Furthermore, $\overline{RT}$ is between each pair of congruent angles listed above. Since every segment is congruent to itself, we also have

$$\overline{RT} \cong \overline{RT}$$

Knowing that two angles and the side between them in △*RST* are congruent, respectively, to two angles and the side between them in △*RUT*, we can conclude that △*RST* ≅ △*RUT* by the ASA property.

Answer: yes, by the SSS property

Similar triangles

We have seen that congruent triangles have the same shape and size. **Similar triangles** have the same shape, but not necessarily the same size. That is, one triangle is an exact scale model of the other triangle. If the triangles in Figure 106 are similar, we can write △*ABC* ~ △*DEF* (read the symbol ~ as "is similar to").

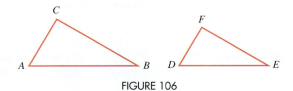

FIGURE 106

   COMMENT   Note that congruent triangles are always similar, but similar triangles are not always congruent.

The formal definition of similar triangles requires that we establish a correspondence between the vertices of the triangles.

Similar triangles

Two triangles are **similar** if and only if their vertices can be matched so that corresponding angles are congruent and the lengths of corresponding sides are proportional.

EXAMPLE 4 *Similar triangles.*   Refer to Figure 107. If △*PQR* ~ △*CDE*, name the congruent angles and the sides that are proportional.

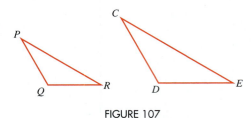

FIGURE 107

Solution

When we write △*PQR* ~ △*CDE,* a correspondence between the vertices of the triangles is established.

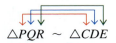

△*PQR* ~ △*CDE*

Since the triangles are similar, corresponding angles are congruent:

$$\angle P \cong \angle C, \quad \angle Q \cong \angle D, \quad \angle R \cong \angle E$$

The lengths of the corresponding sides are proportional. (To simplify the notation, we will now let $PQ = m(\overline{PQ})$, $CD = m(\overline{CD})$, $QR = m(\overline{QR})$, and so on.)

$$\frac{PQ}{CD} = \frac{QR}{DE}, \quad \frac{QR}{DE} = \frac{PR}{CE}, \quad \frac{PQ}{CD} = \frac{PR}{CE}$$

Written in a more compact way,

$$\frac{PQ}{CD} = \frac{QR}{DE} = \frac{PR}{CE}$$

Self Check

If △*GEF* ~ △*IJH,* name the congruent angles and the sides that are proportional.

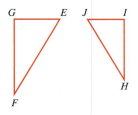

Answers: $\angle G \cong \angle I$, $\angle E \cong \angle J$, $\angle F \cong \angle H$; $\dfrac{EG}{JI} = \dfrac{GF}{IH}$, $\dfrac{GF}{IH} = \dfrac{FE}{HJ}$, $\dfrac{EG}{JI} = \dfrac{FE}{HJ}$ ■

Property of similar triangles	If two triangles are similar, all pairs of corresponding sides are in proportion.

It is possible to conclude that two triangles are similar without having to show that all three pairs of corresponding angles are congruent and that the lengths of all three pairs of corresponding sides are proportional.

AAA similarity theorem	If the angles of one triangle are congruent to corresponding angles of another triangle, the triangles are similar.

EXAMPLE 5 *Determining whether triangles are similar.*   In Figure 108, $\overline{PR} \parallel \overline{MN}$. Are △*PQR* and △*NQM* similar triangles?

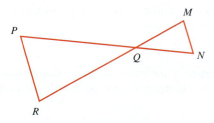

FIGURE 108

Self Check

In the figure below, $\overline{YA} \parallel \overline{ZB}$. Are △*XYA* and △*XZB* similar triangles?

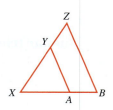

Solution

Since vertical angles are congruent,

$\angle PQR \cong \angle NQM$

In the figure, we can view $\overleftrightarrow{PN}$ as a transversal cutting parallel line segments $\overline{PR}$ and $\overline{MN}$. Since alternate interior angles are congruent, we have

$\angle RPQ \cong \angle MNQ$

Furthermore, we can view $\overleftrightarrow{RM}$ as a transversal cutting parallel line segments $\overline{PR}$ and $\overline{MN}$. Alternate interior angles are congruent.

$\angle QRP \cong \angle QMN$

These observations are summarized in Figure 109.

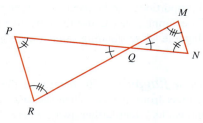

FIGURE 109

In the figure, we see that corresponding angles of $\triangle PQR$ are congruent to corresponding angles of $\triangle NQM$. By the AAA similarity theorem, we can conclude that

$\triangle PQR \sim \triangle NQM$

Answer: yes   ■

EXAMPLE 6   *Finding the length of a side of a triangle.*   In Figure 110, $\triangle RST \sim \triangle JKL$. Find   **a.** x and   **b.** y.

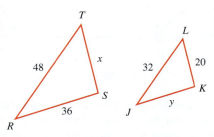

FIGURE 110

Self Check

In the figure below, $\triangle DEF \sim \triangle GHI$. Find   **a.** x and   **b.** y.

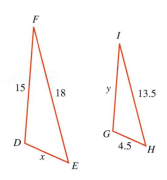

Solution

a. Since $\triangle RST \sim \triangle JKL$, the lengths of corresponding sides of $\triangle RST$ and $\triangle JKL$ are proportional. To find x, we write a proportion of corresponding sides so that x is the only unknown.

$$\frac{RT}{JL} = \frac{ST}{KL}$$ Each fraction is a ratio of a side length of $\triangle RST$ to its corresponding side length of $\triangle JKL$.

$$\frac{48}{32} = \frac{x}{20}$$ Substitute: $RT = 48$, $JL = 32$, $ST = x$, and $KL = 20$.

$48(20) = 32x$ In a proportion, the product of the extremes is equal to the product of the means.

$960 = 32x$ Do the multiplication: $48(20) = 960$.

$30 = x$ To undo the multiplication by 32, divide both sides by 32.

$x = 30$

b. To find *y*, we write a proportion of corresponding side lengths in such a way that *y* is the only unknown.

$$\frac{RT}{JL} = \frac{RS}{JK}$$

$\frac{48}{32} = \frac{36}{y}$ Substitute: $RT = 48$, $JL = 32$, $RS = 36$, and $JK = y$.

$48y = 32(36)$ In a proportion, the product of the extremes is equal to the product of the means.

$48y = 1,152$ Do the multiplication: $32(36) = 1,152$.

$\quad y = 24$ To undo the multiplication by 48, divide both sides by 48.   **Answers: a.** 6, **b.** 11.25 ■

Similar triangles and proportions can be used to find lengths that would normally be difficult to measure. For example, we can use the reflective properties of a mirror to calculate the height of a flagpole while standing safely on the ground.

EXAMPLE 7   ***Finding the height of a flagpole.***   To determine the height of a flagpole, a woman walks to a point 20 feet from its base. Then she takes a mirror from her purse, places it on the ground, and walks 2 feet farther away, where she can see the top of the pole reflected in the mirror. (See Figure 111.) Find the height of the pole.

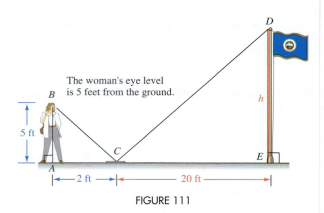

FIGURE 111

Solution   We can solve this problem using similar triangles. To show that $\triangle ABC \sim \triangle EDC$, we begin by applying an important fact about mirrors. When a beam of light strikes a mirror, it is reflected at the same angle as it hits the mirror. Therefore, $\angle BCA \cong \angle DCE$. Furthermore, $\angle A \cong \angle E$, because the woman and the flagpole are perpendicular to the ground. Finally, if two pairs of corresponding angles are congruent, it follows that the third pair of corresponding angles are also congruent: $\angle B \cong \angle D$. By the AAA similarity theorem, we conclude that $\triangle ABC \sim \triangle EDC$.

Since the triangles are similar, the lengths of their corresponding sides are in proportion. If we let *h* represent the height of the flagpole, we can find *h* by solving the following proportion.

$\frac{h}{5} = \frac{20}{2}$ $\frac{\text{Height of the pole}}{\text{Eye level of the woman}} = \frac{\text{Distance the pole is from the mirror}}{\text{Distance the woman is from the mirror}}$

$2h = 5(20)$ In a proportion, the product of the extremes is equal to the product of the means.

$2h = 100$

$\quad h = 50$

The flagpole is 50 feet tall.   ■

STUDY SET   Section 8

VOCABULARY   *Fill in the blanks.*

1. _____ triangles are the same size and the same shape.

2. When we match the vertices of △*ABC* with the vertices of △*DEF*, as shown below, we call this matching of points a _____.

$$A \leftrightarrow D \qquad B \leftrightarrow E \qquad C \leftrightarrow F$$

3. Two angles or two line segments with the same measure are said to be _____.

4. All _____ parts of congruent triangles have the same measure.

5. If two triangles are _____, they have the same shape.

6. A _____ is a mathematical statement that two ratios (fractions) are equal.

CONCEPTS

7. Refer to the triangles in Illustration 1.

ILLUSTRATION 1

  a. Do these triangles appear to be congruent? Explain why or why not.

  b. Do these triangles appear to be similar? Explain why or why not.

8. a. Draw a triangle that is congruent to △*CDE* shown in Illustration 2.

  b. Draw a triangle that is similar to, but not congruent to, △*CDE* shown in Illustration 2.

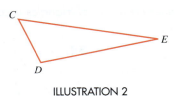

ILLUSTRATION 2

Name the corresponding parts of the congruent triangles.

9. Refer to Illustration 3.

$\overline{AC} \cong$ _____

$\overline{DE} \cong$ _____

$\overline{BC} \cong$ _____

$\angle A \cong$ _____

$\angle E \cong$ _____

$\angle F \cong$ _____

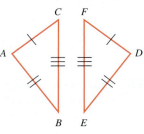

ILLUSTRATION 3

10. Refer to Illustration 4.

$\overline{AB} \cong$ _____

$\overline{EC} \cong$ _____

$\overline{AC} \cong$ _____

$\angle D \cong$ _____

$\angle B \cong$ _____

$\angle 1 \cong$ _____

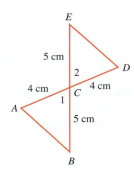

ILLUSTRATION 4

In Exercises 11–14, fill in the blanks.

11. △*XYZ* ≅ △____

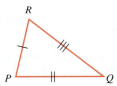

12. △____ ≅ △*DEF*

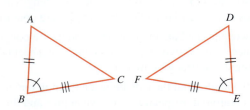

13. △*RST* ~ △____

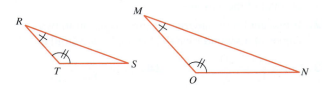

14. △_____ ~ △*TAC*

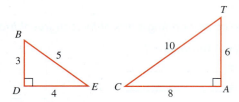

15. Name the six corresponding parts of the congruent triangles shown in Illustration 5.

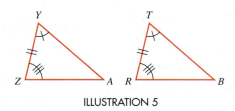

ILLUSTRATION 5

16. Refer to the similar triangles in Illustration 6.
   a. Name 3 pairs of congruent angles.
   b. Complete each proportion.

$$\frac{LM}{HJ} = \frac{}{JE}, \quad \frac{MR}{} = \frac{LR}{HE}, \quad \frac{}{HJ} = \frac{LR}{HE}$$

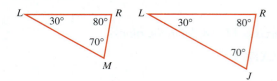

ILLUSTRATION 6

In Exercises 17–22, tell whether each statement is true. If a statement is false, tell why.

17. If three sides of one triangle are the same length as the corresponding three sides of a second triangle, the triangles are congruent.

18. If two sides of one triangle are the same length as two sides of a second triangle, the triangles are congruent.

19. If two sides and an angle of one triangle are congruent, respectively, to two sides and an angle of a second triangle, the triangles are congruent.

20. If two angles and the side between them in one triangle are congruent, respectively, to two angles and the side between them in a second triangle, the triangles are congruent.

21. In a proportion, the product of the means is equal to the product of the extremes.

22. If two angles of one triangle are congruent to two angles of a second triangle, the angles are similar.

23. Solve $\dfrac{x}{15} = \dfrac{20}{3}$.

24. Solve $\dfrac{h}{2.6} = \dfrac{27}{13}$.

25. The symbol ≅ is read as "_____."
26. The symbol ~ is read as "_____."
27. Use markings to show the congruent parts of the triangles shown in Illustration 7.

$$\angle K \cong \angle H \qquad \overline{KR} \cong \overline{HJ} \qquad \angle M \cong \angle E$$

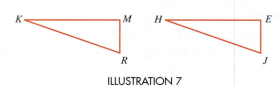

ILLUSTRATION 7

28. Use markings to show the congruent parts of the triangles shown in Illustration 8.

$$\angle P \cong \angle T \qquad \overline{LP} \cong \overline{RT} \qquad \overline{FP} \cong \overline{ST}$$

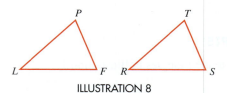

ILLUSTRATION 8

PRACTICE   *Fill in the blanks.*

29. Two triangles are _____ if and only if their vertices can be matched so that the corresponding sides and the corresponding angles are congruent.

30. If three sides of one triangle are congruent to three sides of a second triangle, the triangles are _____.

31. If two sides and the angle between them in one triangle are congruent, respectively, to two sides and the angle between them in a second triangle, the triangles are _____.

32. If two angles and the side between them in one triangle are congruent, respectively, to two angles and the side between them in a second triangle, the triangles are _____.

Determine whether each pair of triangles is congruent. If they are, tell why.

33.   **34.**

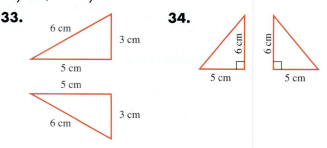

35.

6 m
6 m

36.
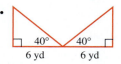

46. △ABC ≅ △DEC
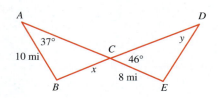
A
37°
10 mi
C
y
46°
x
8 mi
B
E
D

37.

38.
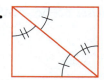
40°
40°

Find x.

47.
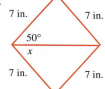
5 mm
6 mm
x mm
5 mm

48.
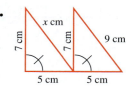
x cm
7 cm
7 cm
9 cm
5 cm   5 cm

39.

40.

40°   40°
6 yd   6 yd

49.
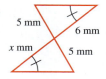
7 in.   7 in.
50°
x
7 in.   7 in.

50.
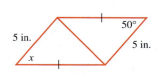
50°
5 in.
5 in.
x

41. $\overline{AB} \parallel \overline{DE}$

42. $\overline{XY} \parallel \overline{ZQ}$

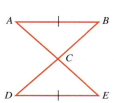
A   B
C
D   E

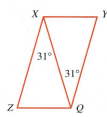
X   Y
31°
31°
Z   Q

Fill in the blanks.

51. Two triangles are similar if and only if their vertices can be matched so that corresponding angles are congruent and the lengths of corresponding sides are _____.

52. If the angles of one triangle are congruent to corresponding angles of another triangle, the triangles are _____.

In Exercises 43–44, △ABC ≅ △DEF. Find x and y.

43.
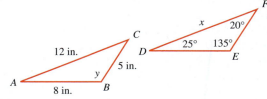
F
x   20°
C
25°   135°
12 in.   D   E
5 in.
y
A   B
8 in.

Tell whether the triangles are similar.

53.
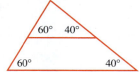
60°   40°
60°   40°

44.
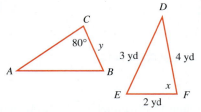
D
C
80°   y
3 yd   4 yd
A   B
x
E   F
2 yd

54.
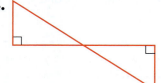

55. 4

6

56.

In Exercises 45–46, find x and y.

45. △ACB ≅ △ADB

57.

70°   40°
40°   70°

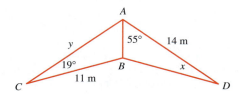
A
y   55°   14 m
19°   B   x
C   11 m   D

58.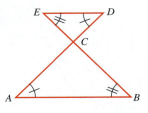

59.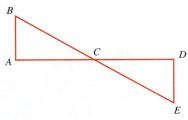

60.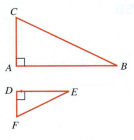

61. $\overline{XY} \parallel \overline{ZD}$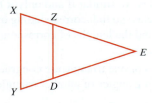

62. $\overline{QR} \parallel \overline{TU}$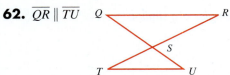

In Exercises 63–64, △MSN ~ △TPR. Find x and y.

63.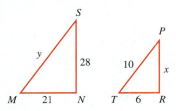

64.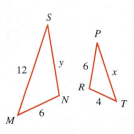

In Exercises 65–66, △MSN ~ △TPN. Find x and y.

65. 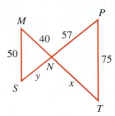   **66.**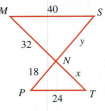

67. If $\overline{DE}$ in Illustration 9 is parallel to $\overline{AB}$, △ABC will be similar to △DEC. Find *x*.

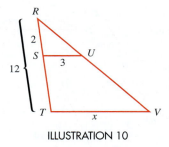

ILLUSTRATION 9

68. If $\overline{SU}$ in Illustration 10 is parallel to $\overline{TV}$, △SRU will be similar to △TRV. Find *x*.

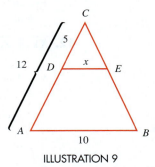

ILLUSTRATION 10

69. If $\overline{DE}$ in Illustration 11 is parallel to $\overline{CB}$, △EAD will be similar to △BAC. Find *x*.

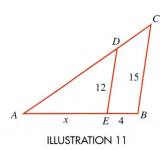

ILLUSTRATION 11

70. If $\overline{HK}$ in Illustration 12 is parallel to $\overline{AB}$, $\triangle HCK$ will be similar to $\triangle ACB$. Find x.

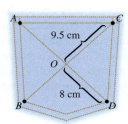

ILLUSTRATION 12

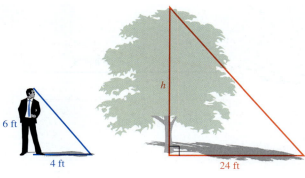

ILLUSTRATION 15

APPLICATIONS

71. SEWING   The pattern that is sewn on the rear pocket of a pair of blue jeans is shown Illustration 13. If $\triangle AOB \cong \triangle COD$, how long is the stitching from point A to point D?

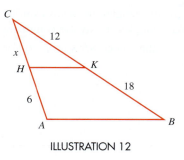

ILLUSTRATION 13

72. CAMPING   Refer to Illustration 14. The base of the tent pole is placed at the midpoint between the stake at point A and the stake at point B, and it is perpendicular to the ground. Explain why $\triangle ACD \cong \triangle BCD$.

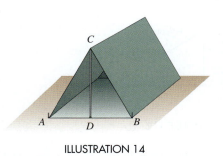

ILLUSTRATION 14

Solve each problem. If an answer is not exact, give the answer to the nearest tenth.

73. HEIGHT OF A TREE   The tree in Illustration 15 casts a shadow 24 feet long when a man 6 feet tall casts a shadow 4 feet long. Find the height of the tree.

74. HEIGHT OF A BUILDING   A man places a mirror on the ground and sees the reflection of the top of a building, as shown in Illustration 16. Find the height of the building.

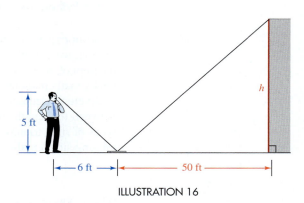

ILLUSTRATION 16

75. WIDTH OF A RIVER   Use the dimensions in Illustration 17 to find w, the width of the river.

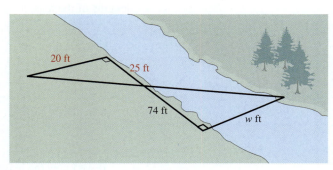

ILLUSTRATION 17

76. FLIGHT PATH   The airplane in Illustration 18 (on the next page) ascends 200 feet as it flies a horizontal distance of 1,000 feet. How much altitude is gained as it flies a horizontal distance of 1 mile? (*Hint:* 1 mile = 5,280 feet.)

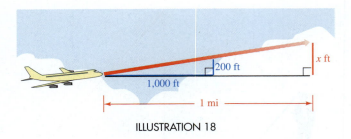

ILLUSTRATION 18

77. Tell whether the statement is true or false. Explain your answer.
  a. Congruent triangles are always similar.
  b. Similar triangles are always congruent.

78. Explain why there is no SAA property for congruent triangles.

9 *The Pythagorean Theorem and Special Triangles*

In this section, you will learn about

• The Pythagorean theorem • 45°–45°–90° triangles • 30°–60°–90° triangles

INTRODUCTION. A **theorem** is a mathematical statement that can be proven. In this section, we will discuss what is probably the most widely known and most often used theorem of geometry—the Pythagorean theorem. It is so named because **Pythagoras,** a Greek mathematician who lived about 2,500 years ago, is thought to have been the first to develop a deductive proof of it. The Pythagorean theorem expresses the relationship between the lengths of the sides of any right triangle. We will also study two specific types of right triangles and we will derive some formulas that can be used to find their missing side lengths.

The Pythagorean theorem

Recall that a **right triangle** is a triangle that has a right angle (an angle with measure 90°). In a right triangle, the longest side is called the **hypotenuse.** It is the side opposite the right angle. The other two sides are called **legs.** It is common practice to let the variable c represent the length of the hypotenuse and the variables a and b represent the lengths of the legs, as shown in Figure 112.

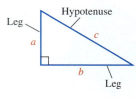

FIGURE 112

If we know the lengths of any two sides of a right triangle, we can find the length of the third side using the **Pythagorean theorem.**

Pythagorean theorem

If a and b represent the lengths of two legs of a right triangle and c represents the length of the hypotenuse, then
$$a^2 + b^2 = c^2$$

In words, the Pythagorean theorem is expressed as follows:

> *In any right triangle, the square of the hypotenuse is equal to the sum of the squares of the two legs.*

Suppose the right triangle shown in Figure 113 has legs of length 3 and 4 units. To find the length of the hypotenuse, we use the Pythagorean theorem.

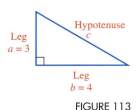

Leg
$a = 3$

Hypotenuse
c

Leg
$b = 4$

FIGURE 113

$$a^2 + b^2 = c^2$$
$$3^2 + 4^2 = c^2 \quad \text{Substitute 3 for } a \text{ and 4 for } b.$$
$$9 + 16 = c^2$$
$$25 = c^2$$

According to the square root property, the equation $25 = c^2$ has two solutions: $c = \sqrt{25}$ and $c = -\sqrt{25}$. Since c represents the length of a side of a triangle, it follows that c is the positive square root of 25.

$$\sqrt{25} = c$$
$$5 = c$$

The length of the hypotenuse is 5 units.

EXAMPLE 1   *Firefighting.*

To fight a forest fire, the forestry department plans to clear a rectangular fire break around the fire, as shown in Figure 114. Crews are equipped with mobile communications that have a 3,000-yard range. Can crews at points *A* and *B* remain in radio contact?

Solution

Points *A*, *B*, and *C* form a right triangle. To find the distance *c* from point *A* to point *B*, we can use the Pythagorean theorem, substituting 2,400 for *a* and 1,000 for *b* and solving for *c*.

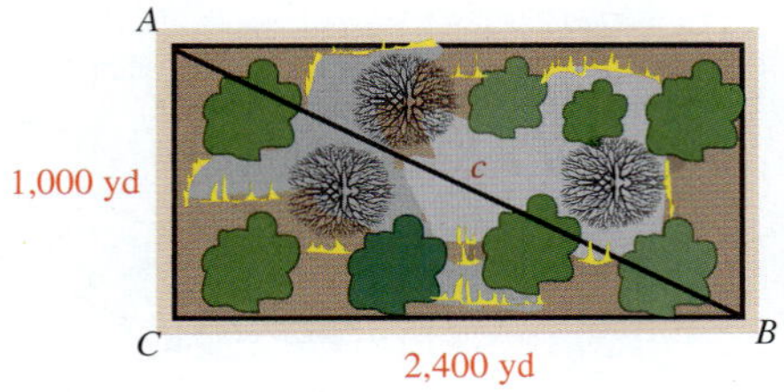

$$a^2 + b^2 = c^2$$
$$2{,}400^2 + 1{,}000^2 = c^2$$
$$5{,}760{,}000 + 1{,}000{,}000 = c^2$$
$$6{,}760{,}000 = c^2$$
$$\sqrt{6{,}760{,}000} = c \quad \begin{array}{l}\text{Apply the square root property: } \sqrt{6{,}760{,}000} = c \\ \text{or } -\sqrt{6{,}760{,}000} = c.\ \text{Since } c \text{ represents a length,} \\ \text{consider only the positive solution.}\end{array}$$
$$2{,}600 = c \quad \text{Use a calculator to find the square root.}$$

The two crews are 2,600 yards apart. Because this distance is less than the range of the radios, they can communicate by radio.

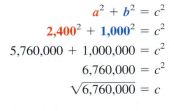

A

1,000 yd

c

C

2,400 yd

B

FIGURE 114

Self Check

In Example 1, can the crews communicate if the line segment $\overline{AC}$ has length 1,500 yards?

Answer: yes

EXAMPLE 2   *Finding the length of a leg of a right triangle.*

Refer to Figure 115. Find the missing length.

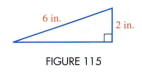

6 in.

2 in.

FIGURE 115

Self Check

Refer to the triangle below. Find the missing length. Give the exact answer and an approximation to the nearest tenth.

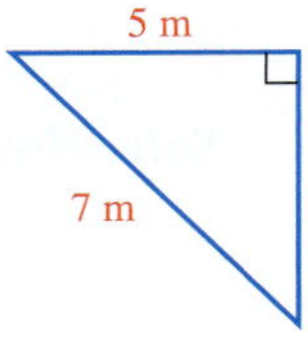

5 m

7 m

Solution

Since the given triangle is a right triangle, we can use the Pythagorean theorem to find the missing length. We may substitute 2 for either *a* or *b*, but 6 must be substituted for the length *c* of the hypotenuse.

$$a^2 + b^2 = c^2$$

$$2^2 + b^2 = 6^2 \qquad \text{Substitute 2 for } a \text{ and 6 for } c.$$

$$4 + b^2 = 36$$

$$4 + b^2 - 4 = 36 - 4 \qquad \text{To isolate } b^2 \text{ on the left-hand side,}$$
$$\text{subtract 4 from both sides.}$$

$$b^2 = 32$$

Now we apply the square root property. Since *b* represents the length of a side of a triangle, we consider only the positive square root.

$$b = \sqrt{32}$$

$$b = 4\sqrt{2} \qquad \text{Simplify the radical expression:}$$
$$\sqrt{32} = \sqrt{16 \cdot 2} = 4\sqrt{2}.$$

The missing side length is exactly $4\sqrt{2}$ inches. We can use a calculator to approximate the length. To the nearest tenth, it is 5.7 inches.

Answer: $2\sqrt{6} \text{ m} \approx 4.9 \text{ m}$ ■

EXAMPLE 3   *Recording ozone levels.*   Every day, a technician drives the route between the three pollution monitoring stations shown in Figure 116. The downtown/east leg of the trip is 3 miles longer than the downtown/west leg. The east/west trip is 6 miles longer than the downtown/west leg. How far apart are the stations?

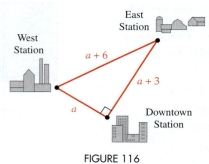

FIGURE 116

Analyze the problem   To find the distances between the stations, we need to find the lengths of the sides of the right triangle formed by connecting their positions. The Pythagorean theorem gives the relationship between the sides of a right triangle: $a^2 + b^2 = c^2$.

Form an equation   We let *a* represent the length of the downtown/west leg of the trip, because the other two distances can be expressed in terms of it. The downtown/east leg is therefore $a + 3$ miles, and the east/west trip is $a + 6$ miles. We substitute these distances into the Pythagorean theorem, noting that the length *c* of the hypotenuse is $a + 6$ miles.

$$a^2 + b^2 = c^2 \qquad \text{The Pythagorean theorem.}$$

$$a^2 + (a + 3)^2 = (a + 6)^2 \qquad \text{Substitute } (a + 3) \text{ for } b \text{ and } (a + 6) \text{ for } c.$$

$$a^2 + a^2 + 6a + 9 = a^2 + 12a + 36 \qquad \text{Find } (a + 3)^2 \text{ and } (a + 6)^2.$$

$$2a^2 + 6a + 9 = a^2 + 12a + 36 \qquad \text{Combine like terms on the left-hand side.}$$

$$a^2 - 6a - 27 = 0 \qquad \text{Subtract } a^2, 12a, \text{ and 36 from both sides to}$$
$$\text{make the right-hand side zero.}$$

Solve the equation   Now we solve the quadratic equation for *a*.

$$a^2 - 6a - 27 = 0$$

$$(a - 9)(a + 3) = 0 \qquad \text{Factor.}$$

$$a - 9 = 0 \quad \text{or} \quad a + 3 = 0 \quad \text{Set each factor equal to zero.}$$
$$a = 9 \qquad\qquad a = -3 \quad \text{Solve each linear equation.}$$

State the conclusion   Since a triangle cannot have a negative number for the length of a side, we discard the result $a = -3$. The downtown/west leg is 9 miles. The downtown/east leg is $9 + 3$, or 12 miles. The distance between the east and west stations is $9 + 6$, or 15 miles.

Check the result   The differences in the distances between stations meet the requirements stated in the problem. The distances also satisfy the Pythagorean theorem. So the results check.

$$9^2 + 12^2 \stackrel{?}{=} 15^2$$
$$81 + 144 \stackrel{?}{=} 225$$
$$225 = 225$$ ∎

Recall that if a mathematical statement is written in the form *if p . . ., then q . . .,* we call the statement *if q . . ., then p . . .* its **converse.** The converses of some statements are true, while the converses of other statements are false. It is interesting to note that the converse of the Pythagorean theorem is true.

Converse of the Pythagorean theorem	If a triangle has three sides of lengths a, b, and c, such that $a^2 + b^2 = c^2$, then the triangle is a right triangle.

EXAMPLE 4   *Determining whether a triangle is a right triangle.*
Is the triangle in Figure 117 a right triangle?

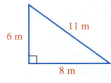

FIGURE 117

Self Check
Is the triangle below a right triangle?

Solution
By the converse of the Pythagorean theorem, the triangle is a right triangle if a true statement results when the side lengths of 6, 8, and 11 are substituted into $a^2 + b^2 = c^2$. We must substitute the longest side length, 11, for c, because it is the possible hypotenuse. The lengths of 6 and 8 may be substituted for either a or b.

$$\boldsymbol{a^2 + b^2 = c^2}$$
$$\boldsymbol{6^2 + 8^2} \stackrel{?}{=} \boldsymbol{11^2}$$
$$36 + 64 \stackrel{?}{=} 121$$
$$100 \ne 121$$

Since $100 \ne 121$, the triangle is not a right triangle.

Answer: no   ∎

45°–45°–90° triangles
An **isosceles right triangle** is a right triangle with two legs of equal length. Isosceles right triangles have angle measures of 45°, 45°, and 90°. If we know the length of one leg of an isosceles right triangle, we can use the Pythagorean theorem to find the length of the hypotenuse. Since the triangle shown in Figure 118 is a right triangle, we have

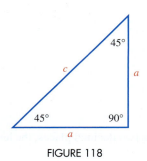

FIGURE 118

$$c^2 = a^2 + b^2$$
$$c^2 = a^2 + a^2 \quad \text{Both legs are } a \text{ units long, so replace } b \text{ with } a.$$
$$c^2 = 2a^2 \quad\quad \text{Combine like terms.}$$

$$c = \sqrt{2a^2} \quad \text{Take the positive square root of both sides.}$$

$$c = a\sqrt{2} \quad \text{Simplify the radical: } \sqrt{2a^2} = \sqrt{2}\sqrt{a^2} = \sqrt{2}a = a\sqrt{2}.$$

This result is a formula that can be used to find the length of the hypotenuse of an isosceles right triangle, given the length of a leg.

Isosceles right triangles

> The length c of the hypotenuse of an isosceles right triangle is $\sqrt{2}$ times the length a of either leg:
>
> $$c = a\sqrt{2}$$

 COMMENT   The formula $c = a\sqrt{2}$ may also be written as $c = \sqrt{2}a$. However, we must be very careful when writing it that way, because only the 2 (*not* the a) is under the radical symbol.

EXAMPLE 5   *45°–45°–90° triangles.*   If one leg of an isosceles right triangle is 10 feet long, find the length of the hypotenuse.

Solution
Figure 119 shows what we are given, and what we are to find. Since the length of the hypotenuse is the length of a leg times $\sqrt{2}$, we have

$$c = a\sqrt{2}$$
$$c = 10\sqrt{2} \quad \text{Substitute 10 for } a.$$

The exact length of the hypotenuse is $10\sqrt{2}$ feet. To two decimal places, the length is 14.14 feet.

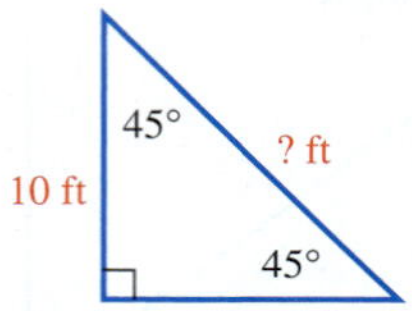

Self Check

Find the length of the hypotenuse of an isosceles right triangle if one leg is 12 meters long.

FIGURE 119

Answer: $12\sqrt{2}$ m ≈ 16.97 m

EXAMPLE 6   *Finding the length of each leg.*   Find the exact length and an approximation of the length of each leg of the triangle shown in Figure 120.

Solution   We will let a represent the unknown length of each leg of the isosceles right triangle. From the figure, we see that the length of the hypotenuse is 25 cm. To find the length of a leg, we substitute 25 for c in the formula $c = a\sqrt{2}$ and solve for a.

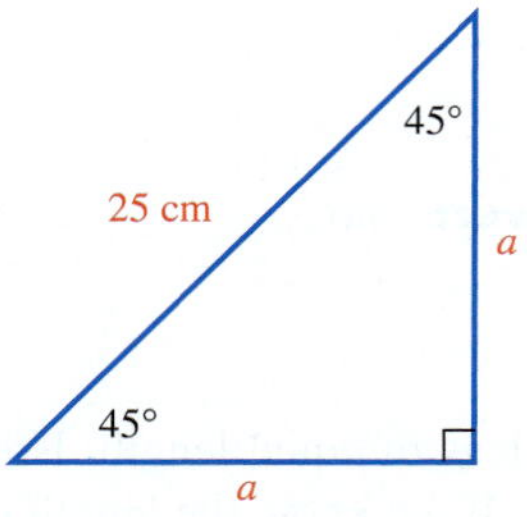

FIGURE 120

$$c = a\sqrt{2}$$
$$25 = a\sqrt{2}$$
$$\frac{25}{\sqrt{2}} = \frac{a\sqrt{2}}{\sqrt{2}} \quad \text{To undo the multiplication by } \sqrt{2}, \text{ divide both sides by } \sqrt{2}.$$
$$\frac{25}{\sqrt{2}} = a$$
$$\frac{25 \cdot \sqrt{2}}{\sqrt{2} \cdot \sqrt{2}} = a \quad \text{Rationalize the denominator.}$$
$$\frac{25\sqrt{2}}{2} = a \quad \text{Simplify: } \sqrt{2} \cdot \sqrt{2} = 2.$$

The exact length of each leg of the triangle is $\frac{25\sqrt{2}}{2}$ cm. To two decimal places, the length is 17.68 cm.

30°–60°–90° triangles

Recall that an **equilateral triangle** is a triangle with three sides of equal length and three 60° angles. Each side of the equilateral triangle in Figure 121 is $2a$ units long. If an **altitude** (shown in green) is drawn to its base, the altitude bisects the base and divides the equilateral triangle into two congruent 30°–60°–90° triangles. We see that the shorter leg of each 30°–60°–90° triangle (the side *opposite* the 30° angle) is a units long. That is, it is half as long as the hypotenuse. Another way to express this relationship is that the hypotenuse is twice as long as the shorter leg.

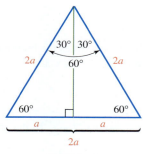

FIGURE 121

 COMMENT   In a 30°–60°–90° triangle, the shorter leg is opposite the 30° angle, and the longer leg is opposite the 60° angle.

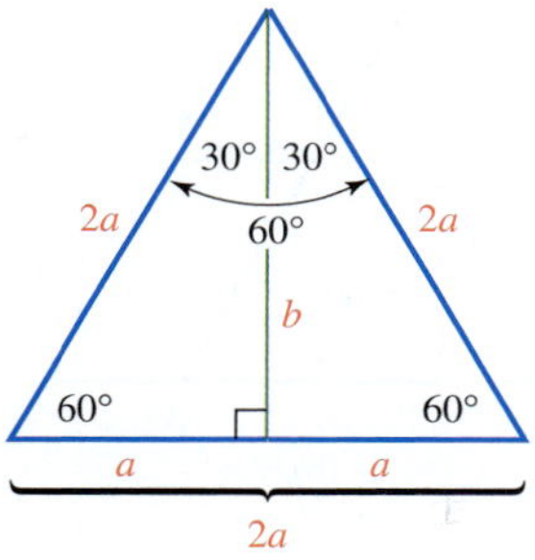

We can discover another important relationship between the legs of a 30°–60°–90° triangle if we find the length b of the altitude in Figure 122. We begin by applying the Pythagorean theorem to one of the 30°–60°–90° triangles.

FIGURE 122

$$a^2 + b^2 = c^2$$

$$a^2 + b^2 = (2a)^2 \quad \text{Since the hypotenuse is } 2a \text{ units long, replace } c \text{ with } 2a.$$

$$a^2 + b^2 = 4a^2 \quad \text{Find the 2nd power of } 2a: (2a)^2 = (2a)(2a) = 4a^2.$$

$$b^2 = 3a^2 \quad \text{To isolate } b^2, \text{ subtract } a^2 \text{ from both sides: } 4a^2 - a^2 = 3a^2.$$

$$b = \sqrt{3a^2} \quad \text{Apply the square root property. Since } b \text{ represents the length of a side of a triangle, we discard the solution } b = -\sqrt{3a^2}.$$

$$b = a\sqrt{3} \quad \text{Simplify the radical: } \sqrt{3a^2} = \sqrt{3}\sqrt{a^2} = \sqrt{3}a = a\sqrt{3}.$$

We see that the longer leg of the 30°–60°–90° triangle is $\sqrt{3}$ times as long as the shorter leg.

 COMMENT   The formula $b = a\sqrt{3}$ may also be written as $b = \sqrt{3}a$. However, we must be very careful when writing it that way, because only the 3 (not the a) is under the radical symbol.

30°–60°–90° triangles

> For any 30°–60°–90° triangle,
>
> **1.** The length c of the hypotenuse is twice the length a of the shorter leg:
> $$c = 2a$$
> **2.** The length b of the longer leg is $\sqrt{3}$ times the length a of the shorter leg:
> $$b = a\sqrt{3}$$

EXAMPLE 7 *30°–60°–90° triangles.*   Find the length of the hypotenuse and the longer leg of the right triangle shown in Figure 123.

Solution

For any 30°–60°–90° triangle, the hypotenuse is twice as long as the shorter leg. We can find the length c of the hypotenuse by substituting 6 for a in the formula $c = 2a$.

$$c = 2a$$

$$c = 2(6) \quad \text{The length of the shorter leg is 6 cm.}$$

$$= 12$$

The hypotenuse is 12 centimeters long.
The length b of the longer leg is $\sqrt{3}$ times the length of the shorter leg. We can find b by substituting 6 for a in the formula $b = a\sqrt{3}$.

$$b = a\sqrt{3}$$

$$b = 6\sqrt{3}$$

The longer leg is exactly $6\sqrt{3}$ centimeters long. To the nearest hundredth, this is 10.39 centimeters.

FIGURE 123

Self Check

Find the length of the hypotenuse and the longer leg of a 30°–60°–90° triangle if the shorter leg is 8 centimeters long. Approximate any exact answers that contain a radical.

Answers: 16 cm,
$8\sqrt{3}$ cm ≈ 13.86 cm   ■

EXAMPLE 8   *Stretching exercises.*   A doctor prescribed the back-strengthening exercise shown in Figure 124(a) for a patient. The patient was instructed to raise his leg to an angle of 60° and hold the position for 10 seconds. If the patient's leg is 36 inches long, how high off the floor will his foot be when his leg is held at the proper angle?

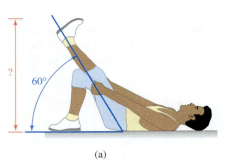
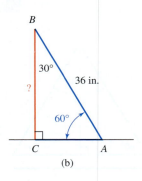

(a)   (b)

FIGURE 124

Solution   In Figure 124(b), we see that a 30°–60°–90° triangle, which we will call △*ABC*, models the situation. Since the side opposite the 30° angle of a 30°–60°–90° triangle is half as long as the hypotenuse, side $\overline{CA}$ is 18 inches long.
Since the length of the side opposite the 60° angle is the length of the side opposite the 30° angle times $\sqrt{3}$, side $\overline{BC}$ is $18\sqrt{3}$, or about 31 inches long. So the patient's foot will be about 31 inches from the floor when his leg is in the proper stretching position.   ■

EXAMPLE 9 *Given the length of the longer leg.*   Find the missing side lengths of the triangle in Figure 125.

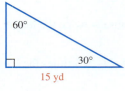

FIGURE 125

Self Check

Find the missing side lengths of the triangle on the next page. Round your answers to the nearest hundredth.

Solution

From the figure, the length of the longer leg is 15 yards. We can find the length a of the shorter leg by substituting 15 for b in the formula $b = a\sqrt{3}$, and solving for a.

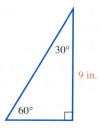

30°

9 in.

60°

$$b = a\sqrt{3}$$

$15 = a\sqrt{3}$ The longer leg is 15 yards long.

$\dfrac{15}{\sqrt{3}} = \dfrac{a\sqrt{3}}{\sqrt{3}}$ To undo the multiplication by $\sqrt{3}$, divide both sides by $\sqrt{3}$.

$\dfrac{15}{\sqrt{3}} = a$

$\dfrac{15 \cdot \sqrt{3}}{\sqrt{3} \cdot \sqrt{3}} = a$ Rationalize the denominator.

$\dfrac{15\sqrt{3}}{3} = a$ Simplify: $\sqrt{3} \cdot \sqrt{3} = 3$.

$5\sqrt{3} = a$ Simplify: $\dfrac{15}{3} = 5$.

The length of the shorter leg is exactly $5\sqrt{3}$ yards. This is about 8.66 yards.

The length of the hypotenuse is $2 \cdot 5\sqrt{3}$ yards, or $10\sqrt{3}$ yards. To the nearest hundredth, this is 17.32 yards.

Answers: $3\sqrt{3}$ in. $\approx$ 5.20 in., $6\sqrt{3}$ in. $\approx$ 10.39 in.   ■

STUDY SET   Section 9

1. In a right triangle, the side opposite the 90° angle is called the _____. The other two sides are called _____.

2. A _____ is a mathematical statement that can be proven.

3. The Pythagorean theorem is named after the Greek mathematician, _____, who is thought to have been the first to prove it.

4. The _____ theorem states that in any right triangle, the square of the length of the hypotenuse is equal to the sum of the squares of the lengths of the two legs.

5. The _____ of the statement *if p ..., then q* is the statement *if q .., then p.*

6. An _____ right triangle is a right triangle with two legs of equal length.

7. An _____ triangle has three sides of equal length and three 60° angles.

8. To solve the _____ equation $a^2 - 6a - 27 = 0$, we begin by factoring $a^2 - 6a - 27$.

CONCEPTS   *Fill in the blanks.*

9. If a and b are the lengths of two legs of a right triangle and c is the length of the hypotenuse, then _____.

10. In any right triangle, the square of the length of the hypotenuse is equal to the _____ of the squares of the lengths of the two _____.

11. The converse of the Pythagorean theorem states: If a triangle has three sides of lengths a, b, and c, such that $a^2 + b^2 = c^2$, then the triangle is a _____ triangle.

12. In a 30°–60°–90° triangle, the shorter leg is opposite the _____ angle, and the longer leg is opposite the _____ angle.

13. In a 30°–60°–90° triangle, the shorter leg is _____ as long as the hypotenuse. Or put another way, the hypotenuse is _____ as long as the shorter leg.

14. In an isosceles right triangle, the length of the hypotenuse is the length of one leg times ____ .

15. The length of the longer leg of a 30°–60°–90° triangle is the length of the shorter leg times ____ .

16. The two solutions of $c^2 = 7$ are $c =$ ____ or $c =$ ____ . If c represents the length of the hypotenuse of a right triangle, then we can discard the solution ____ .

17. What is the first step when solving the equation $25 + b^2 = 81$?

18. Use a protractor to draw an example of each type of triangle.

    a. A right triangle

    b. A right isosceles triangle

    c. A 30°–60°–90° triangle

19. Refer to the triangle in Illustration 1.

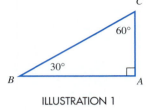

ILLUSTRATION 1

    a. What side is the hypotenuse?

    b. What side is the longer leg?

    c. What side is the shorter leg?

20. When the lengths of the sides of the triangle in Illustration 2 are substituted into the equation $a^2 + b^2 = c^2$, the result is a false statement. Explain why.

$$a^2 + b^2 = c^2$$
$$2^2 + 4^2 = 5^2$$
$$4 + 16 = 25$$
$$20 = 25$$

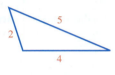

ILLUSTRATION 2

21. Is a triangle with sides of length 12 feet, 14 feet, and 15 feet a right triangle?

22. Approximate each number to the nearest tenth.

    a. $18\sqrt{2}$ **b.** $\dfrac{7\sqrt{3}}{3}$

23. When asked to approximate an exact answer to two decimal places, to what place value column do we round?

24. Simplify each radical expression.

    a. $\sqrt{81}$ **b.** $\sqrt{8}$ **c.** $2\sqrt{27}$

25. Do the multiplication.

    a. $2 \cdot 5\sqrt{2}$ **b.** $\sqrt{3} \cdot \sqrt{3}$

26. Rationalize the denominator.

    a. $\dfrac{16}{\sqrt{2}}$ **b.** $\dfrac{4}{\sqrt{3}}$

NOTATION *Complete the solution to solve the equation.*

27. Solve $8^2 + 4^2 = c^2$, where $c > 0$.

$$\boxed{} + 16 = c^2$$
$$\boxed{} = c^2$$
$$\sqrt{\boxed{}} = \sqrt{c^2}$$
$$\sqrt{\boxed{}} \cdot 5 = c$$
$$\boxed{}\sqrt{5} = c$$
$$c \approx 8.94$$

28. Solve $45 = a\sqrt{3}$.

$$\frac{45}{} = \frac{a\sqrt{3}}{}$$

$$\frac{}{\sqrt{3}} = a$$

$$\frac{45 \cdot }{\sqrt{3} \cdot } = a$$

$$\frac{45\sqrt{3}}{} = a$$

$$\boxed{}\sqrt{3} = a$$

$$a \approx 25.98$$

29. What does $16\sqrt{2}$ mean?

30. True or false: $\sqrt{2} \cdot 7 = 7\sqrt{2}$?

31. Write each formula.

    a. For a 30°–60°–90° triangle, the length b of the longer leg is $\sqrt{3}$ times the length a of the shorter leg.

    b. For a 30°–60°–90° triangle, the length c of the hypotenuse is twice the length a of the shorter leg.

    c. The length c of the hypotenuse of an isosceles right triangle is $\sqrt{2}$ times the length a of either leg.

32. When writing the formula $b = a\sqrt{3}$ as $b = \sqrt{3}a$, is a under the radical symbol?

PRACTICE *The lengths of two sides of the right triangle ABC shown in Illustration 3 are given. Find the length of the missing side. Give the exact answer and then an approximation to one decimal place, when applicable.*

33. $a = 6$ ft and $b = 8$ ft

34. $a = 10$ cm and $c = 26$ cm

35. $b = 18$ m and $c = 82$ m

36. $a = 14$ in. and $c = 50$ in.

37. $a = 5$ cm and $b = 7$ cm

38. $a = 12$ m and $b = 8$ m

39. $a = \sqrt{15}$ in. and $b = 6$ in.

40. $a = 2\sqrt{2}$ yd. and $b = 9$ yd

41. $b = 3\sqrt{3}$ ft and $c = 5\sqrt{2}$ ft

42. $a = 6\sqrt{2}$ mi and $c = 10$ mi

ILLUSTRATION 3

Find x. Then give the lengths of the other two sides of the right triangle.

43.

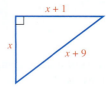

44.

45. In a right triangle, one leg is 7 feet shorter than the other leg. The hypotenuse is 2 feet longer than the longer leg. Find the lengths of the sides of the triangle.

46. In a right triangle, one leg is 3 feet longer than the other. The hypotenuse is 3 feet longer than the longer leg. Find the lengths of the sides of the triangle.

Find the missing lengths in each triangle. Give the exact answer and then an approximation to two decimal places, when applicable.

47.

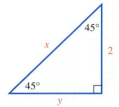

48.

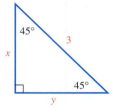

49.

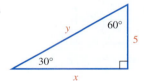

50.

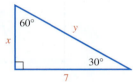

In Exercises 51–54, find the missing lengths in each triangle. Give the answer to two decimal places.

51.

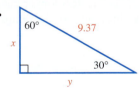

52.

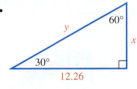

53.

54.

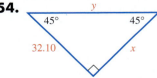

55. Find the exact length of the diagonal (in blue) of one of the *faces* of the cube shown in Illustration 4.

56. Find the exact length of the diagonal (in green) of the cube shown in Illustration 4.

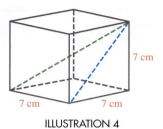

ILLUSTRATION 4

APPLICATIONS   *In Exercises 57–63, give the exact answer. Then give an approximation to two decimal places.*

57. WASHINGTON, DC   The square in Illustration 5 shows the 100-square-mile site selected by George Washington in 1790 to serve as a permanent capital for the United States. In 1847, the part of the district lying on the west bank of the Potomac was returned to Virginia. Find the coordinates of each corner of the original square that outlined the District of Columbia.

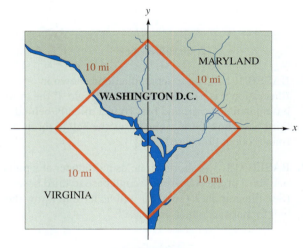

ILLUSTRATION 5

58. PAPER AIRPLANE   Illustration 6 gives the directions for making a paper airplane from a square piece of paper with sides 8 inches long. Find the length *l* of the plane when it is completed.

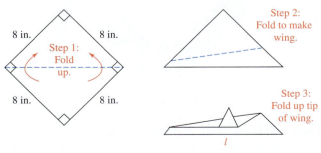

ILLUSTRATION 6

59. HARDWARE   The sides of the regular hexagonal nut shown in Illustration 7 are 10 millimeters long. Find the height *h* of the nut.

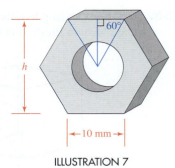

ILLUSTRATION 7

60. IRONING BOARD   Find the height *h* of the ironing board shown in Illustration 8.

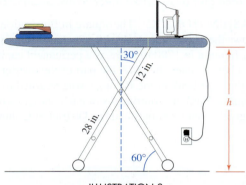

ILLUSTRATION 8

61. BASEBALL   The baseball diamond shown in Illustration 9 is a square, 90 feet on a side. If the third baseman fields a ground ball 10 feet directly behind third base, how far must he throw the ball to throw a runner out at first base?

62. BASEBALL   A shortstop fields a grounder at a point one-third of the way from second base to third base. (See Illustration 9.) How far will he have to throw the ball to make an out at first base?

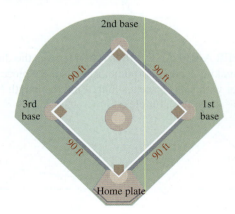

ILLUSTRATION 9

63. CLOTHESLINE   A pair of damp jeans are hung on a clothesline to dry, as shown in Illustration 10. They pull the center down 1 foot. By how much is the line stretched?

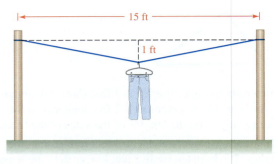

ILLUSTRATION 10

64. FIREFIGHTING   The base of the 37-foot ladder in Illustration 11 is 9 feet from the wall. Will the top reach a window ledge that is 35 feet above the ground? Explain how you arrived at your answer.

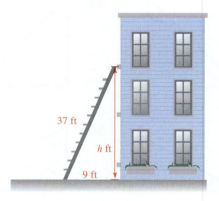

ILLUSTRATION 11

65. BOATING   The inclined ramp of the boat launch shown in Illustration 12 is 8 meters longer than the "rise" of the ramp. The "run" is 7 meters longer than the "rise." How long are the three sides of the ramp?

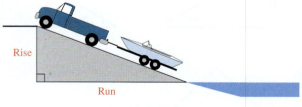

ILLUSTRATION 12

66. GARDENING TOOLS   The dimensions (in millimeters) of the teeth of a pruning saw blade are given in Illustration 13. Find each length.

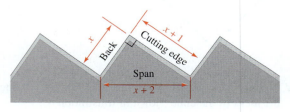

ILLUSTRATION 13

WRITING

67. State the Pythagorean theorem in your own words.

68. In Illustration 14, equal-sized squares have been drawn on the sides of right triangle $\triangle ABC$. Explain how this figure demonstrates that $3^2 + 4^2 = 5^2$.

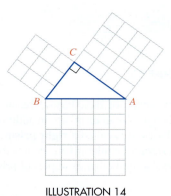

ILLUSTRATION 14

69. In the movie *The Wizard of Oz,* the scarecrow was in search of a brain. To prove that he had found one, he recited the following:

"The sum of the square roots of any two sides of an isosceles triangle is equal to the square root of the remaining side."

Unfortunately, this statement is not true. Correct it so that it states the Pythagorean theorem.

70. List the facts you learned about special right triangles in this section.

10 *Volume*

In this section, you will learn about

• Space figures • Volume • Volume formulas • Volumes of prisms and pyramids • Volumes of cylinders, cones, and spheres

INTRODUCTION. We have studied ways to calculate the perimeter and the area of two-dimensional figures that lie in a plane, such as rectangles, triangles, and circles. Now we will consider three-dimensional figures that occupy space, such as prisms, cylinders, and spheres. In this section, we will introduce the vocabulary associated with these figures, as well as the formulas that are used to find their volume. Volumes are measured in cubic units, such as cubic feet, cubic yards, or cubic centimeters. For example,

- We measure the capacity of a refrigerator in cubic feet.
- We buy gravel or topsoil by the cubic yard.
- We often measure amounts of medicine in cubic centimeters.

Space figures

In geometry, **space** is defined to be the set of all points. **Space figures** are geometric figures that contain points in more than one plane. One example of a space figure is the **prism.** To construct a prism, we begin with two congruent polygons lying in two parallel planes. (See Figure 126 on the next page.) Each polygon and its interior is called a **base** of the prism. The sides of the bases are called **base edges.** When we join the corresponding vertices of each polygon with parallel line segments called **lateral edges,** parallelogram-shaped regions called **lateral faces** are created. A prism is a combination of its bases and its lateral faces. Its **height** h is the distance between the planes that contain its bases.

COMMENT   There are always parts of a three-dimensional figure that cannot be seen from the position of the observer, because they are covered by portions of the figure. The edges of these hidden parts are indicated by dashed lines in a drawing.

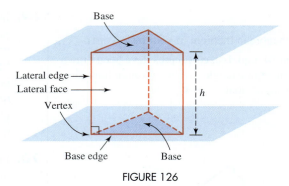

FIGURE 126

Prisms are classified according to the shape of their bases. Because the prism in Figure 126 has a three-sided base, it is called a *triangular* prism. In addition, if the lateral edges of a prism are perpendicular to its bases, it is called a **right prism.** Therefore, the prism in Figure 126 is more specifically a *right* triangular prism. Prisms that are not right prisms are called **oblique.** Figure 127 shows several other examples of prisms.

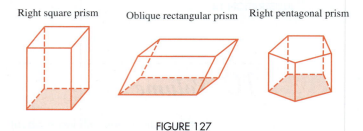

FIGURE 127

 COMMENT   A right rectangular prism (a prism that has two bases and four lateral faces that are rectangles) is commonly referred to as a **rectangular solid.** This name is misleading, because a prism is not solid all the way through like a brick. Visualize a prism as an empty shoe box. If the bases and lateral faces of a prism are squares, it is called a **cube.** Examples of these two special types of prisms are shown in Figure 128.

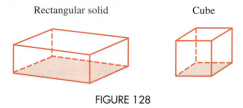

FIGURE 128

Another type of space figure is the **circular cylinder.** Circular cylinders are constructed in a manner similar to prisms. However, the bases of cylinders are congruent circles (circles with the same area). If the segment joining the centers of the circular bases is perpendicular to the bases, the figure is a **right circular cylinder.** Cylinders that are not right cylinders are referred to as **oblique.** The **height** h of a circular cylinder is the distance between the planes that contain the bases. Two examples of circular cylinders are shown in Figure 129.

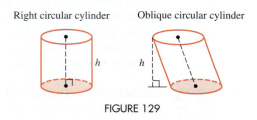

FIGURE 129

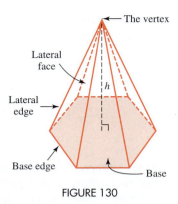

The vertex

Lateral
face

Lateral
edge

h

Base edge

Base

FIGURE 130

To construct another type of space figure, called a **pyramid,** we begin with a polygonal region (the **base**) and a point not in the plane of the region (the **vertex**), as shown in Figure 130. Line segments join the vertex of the pyramid to the vertices of the base. These segments are called **lateral edges.** Each triangular region determined by the edge of the base and two lateral edges is called a **lateral face.** A pyramid is a combination of its base and its lateral faces. Its **height** *h* is the perpendicular distance from the vertex to the plane of the base.

Pyramids are classified in the same way as prisms—by the shape of the base, and as either right or oblique. The vertex of a **right pyramid** is directly above the center of the base. A pyramid whose base is a regular polygon and whose vertex is equidistant from each vertex of the base is called a **regular pyramid.** Therefore, the pyramid in Figure 130 with a 6-sided base is a *right regular* hexagonal pyramid. Two other examples of pyramids are shown in Figure 131.

Right square pyramid

Oblique triangular pyramid

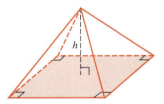

h

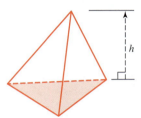

h

FIGURE 131

EXAMPLE 1   *Vocabulary.*   Refer to Figure 132.
a. How many bases does the figure have? What shape are they?
b. How many lateral faces does it have? What shape are they?
c. What is the specific name of the figure?

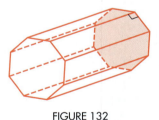

FIGURE 132

Solution
a. This figure has two bases—one directly facing the observer and one at the back that is shaded and partially outlined with dashed line segments. Each base has 8 sides; they are octagons.
b. The figure has 8 lateral faces. Each is a rectangle.
c. Because the lateral edges are perpendicular to the bases, and because the bases have 8 sides, this is a right octagonal prism.

Self Check
Refer to the figure below.

a. How many bases does the figure have? What shape is the base?
b. How many lateral faces does the figure have? What shape are they?
c. What is the specific name of the figure?

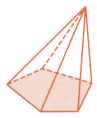

Answers: a. 1, pentagon; **b.** 5, triangles; **c.** oblique pentagonal pyramid ◼

To construct a **circular cone,** we begin with a circle in one plane and then choose a point, called the **vertex,** that is not in the plane. The circular region, together with the set of all segments connecting the vertex to a point on the circle, forms the cone. The variable *r* is normally used to represent the **radius** of the base, and the **height** *h* is the perpendicular distance from the vertex to the plane of the base. If a segment from the vertex of the cone to the center of the base is perpendicular to the base, the cone is called a **right circular cone.** Otherwise, it is said to be **oblique.** Figure 133 shows two examples of circular cones.

Right circular cone   Oblique circular cone

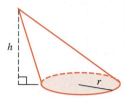

FIGURE 133

The final type of space figure that we will discuss is the **sphere.** A sphere is the set of all points in space that are a given distance, called the **radius,** from a given point, called the **center.** (See Figure 134.) The radius of a sphere is usually represented by the variable *r.*

FIGURE 134

Volume

The **volume** of a three-dimensional figure is a measure of its capacity. Figure 135 shows two common units of volume: cubic inches (in.^{3}) and cubic centimeters (cm^3).

1 cubic inch (1 in.^{3})

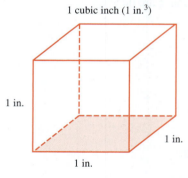

1 cubic centimeter (1 cm^3)

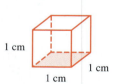

FIGURE 135

The volume of a figure can be thought of as the number of cubic units that will fit within its boundaries. If we divide the right rectangular prism (shown in black) in Figure 136 into cubes, each cube represents a volume of 1 cm^3. Because there are 2 levels with 12 cubes on each level, the volume of the prism is 24 cm^3.

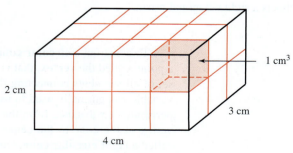

FIGURE 136

Volume formulas

In practice, we do not find volumes by counting cubes. Instead, we use the formulas shown in Table 2. Note that several of the volume formulas involve the variable *B*. It represents the area of the base of the figure.

Cube	Rectangular solid	Sphere
	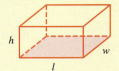	
$V = s^3$	$V = lwh$	$V = \dfrac{4}{3}\pi r^3$
where *s* is the length of a side	where *l* is the length, *w* is the width, and *h* is the height	where *r* is the radius

Prism	Pyramid
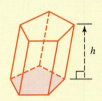	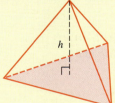
$V = Bh$	$V = \dfrac{1}{3}Bh$
where *B* is the area of the base and *h* is the height	where *B* is the area of the base and *h* is the height

Circular cylinder	Cone
 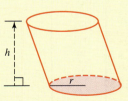	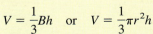
$V = Bh$ or $V = \pi r^2 h$	$V = \dfrac{1}{3}Bh$ or $V = \dfrac{1}{3}\pi r^2 h$
where *B* is the area of the base, *h* is the height, and *r* is the radius	where *B* is the area of the base, *h* is the height, and *r* is the radius

TABLE 2

 COMMENT   The height of a geometric solid is always measured along a line perpendicular to its base.

Volumes of prisms and pyramids

EXAMPLE 2   *Number of cubic inches in one cubic foot.*   How many cubic inches are there in 1 cubic foot? See Figure 137.

Solution

Since a cubic foot is a cube with each side measuring 1 foot, each side also measures 12 inches. Thus, the volume in cubic inches is

$V = s^3$ The formula for the volume of a cube.

$V = (12)^3$ Substitute 12 for s.

$\quad = 1,728$

There are 1,728 cubic inches in 1 cubic foot.

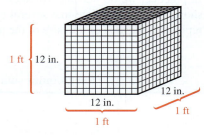

FIGURE 137

Self Check

How many cubic centimeters are in 1 cubic meter?

Answer: 1,000,000 cm^3 ■

EXAMPLE 3   *Volume of an oil storage tank.*   An oil storage tank is in the form of a rectangular solid with dimensions of 17 feet by 10 feet by 8 feet. (See Figure 138.) Find its volume.

Solution

To find the volume, we substitute 17 for l, 10 for w, and 8 for h in the formula $V = lwh$ and simplify.

$V = lwh$

$V = 17(10)(8)$

$\quad = 1,360$

The volume is 1,360 ft^3.

8 ft

17 ft

10 ft

FIGURE 138

Self Check

Find the volume of a rectangular solid with dimensions of 8 meters by 12 meters by 20 meters.

Answer: 1,920 m^3 ■

EXAMPLE 4   *Volume of a triangular prism.*

Find the volume of the triangular prism in Figure 139.

Solution

The volume of the prism is the area of its base multiplied by its height. Since there are 100 centimeters in 1 meter, the height in centimeters is

$$0.5 \text{ m} = 0.5\text{m} \cdot \frac{\textbf{100 cm}}{\textbf{1 m}} \quad \text{Multiply by 1: } \frac{100 \text{ cm}}{1 \text{ m}} = 1.$$

$$\quad = 0.5(\textbf{100 cm})$$

$$\quad = 50 \text{ cm}$$

The area of the triangular base is $\frac{1}{2}(6)(8) = 24$ square centimeters. The height of the prism is 50 centimeters. Substituting into the formula for the volume of a prism, we have

$V = Bh$

$V = 24(50)$

$\quad = 1,200$

The volume of the prism is 1,200 cm^3.

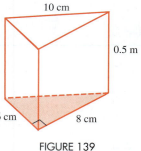

FIGURE 139

Self Check

Find the volume of the triangular prism below.

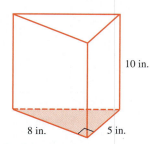

Answer: 200 in.^{3} ■

EXAMPLE 5 *Volume of a pyramid.*   Find the volume of the pyramid shown in Figure 140.

Self Check

Find the volume of the pyramid shown below.

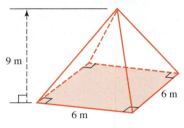

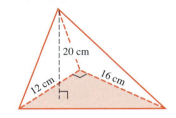

FIGURE 140

Solution

Since the base is a square with each side 6 meters long, the area of the base is $(6 \text{ m})^2$, or 36 m^2. We can then substitute 36 for the area of the base and 9 for the height in the formula for the volume of a pyramid.

$$V = \frac{1}{3}Bh$$

$$V = \frac{1}{3}(36)(9)$$

$= 12(9)$ Multiply: $\frac{1}{3}(36) = \frac{36}{3} = 12.$

$= 108$

The volume of the pyramid is 108 m^3.

Answer: 640 cm^3 ■

EXAMPLE 6   *Egyptian pyramids.*   The largest of all pyramids built by the Egyptians was the Great Pyramid. Constructed about 4,000 years ago near Cairo, it is the only one of the Seven Wonders of the Ancient World still in existence. Its square base covers an area of $571,536 \text{ ft}^2$, and its volume is an astounding $91,636,272 \text{ ft}^3$. Find its height.

Solution   To find the height of the pyramid, we substitute 91,636,272 for V and 571,536 for B in the formula for the volume of a pyramid and then solve for h.

$$V = \frac{1}{3}Bh$$

$$91{,}636{,}272 = \frac{1}{3}(571{,}536)h$$

$$3 \cdot 91{,}636{,}272 = 3 \cdot \frac{1}{3}(571{,}536)h$$ To clear the equation of the fraction, multiply both sides by 3.

$$274{,}908{,}816 = 571{,}536h$$ Multiply: $3 \cdot \frac{1}{3} = 1.$

$$\frac{274{,}908{,}816}{571{,}536} = \frac{571{,}536h}{571{,}536}$$ To undo the multiplication by 571,536, divide both sides by 571,536.

$$481 = h$$ Use a calculator to do the division.

The height of the Great Pyramid is 481 ft.   ■

Volumes of cylinders, cones, and spheres

EXAMPLE 7   Find the volume of the cylinder in Figure 141.

Solution   Since a radius is one-half of the diameter of the circular base, $r = 3$ cm. From the figure, we see that the height of the cylinder is 10 cm. So we substitute 3 for r and 10 for h in the formula for the volume of a cylinder.

$$V = \pi r^2 h$$
$$V = \pi(3)^2(10)$$
$$V = 90\pi \qquad \text{Simplify: } (3)^2(10) = 90.$$
$$\approx 282.7433388 \qquad \text{Use a calculator to do the multiplication.}$$

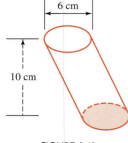

6 cm

10 cm

FIGURE 141

The exact volume of the cylinder is 90π cm^3. To the nearest hundredth, the volume is 282.74 cm^3. ■

**EXAMPLE 8   *Volume of a cone.*   To the nearest tenth, find the volume of the cone in Figure 142.

Solution

Since the radius is one-half of the diameter, $r = 4$ ft. We then substitute 4 for r and 6 for h in the formula for the volume of a cone.

$$V = \frac{1}{3}\pi r^2 h$$

$$V = \frac{1}{3}\pi(4)^2(6)$$

$$V = 32\pi \qquad \text{Find the power: } (4)^2 = 16. \text{ Then multiply: } \frac{1}{3}(6) = 2 \text{ and } 2(16) = 32.$$

$$\approx 100.5309649 \qquad \text{Use a calculator to do the multiplication.}$$

The exact volume of the cone is 32π ft^3. To the nearest tenth, the volume is 100.5 ft^3.

6 ft

8 ft

FIGURE 142

Self Check

Find the volume of the cone shown below.

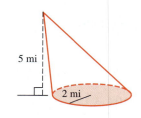

5 mi

2 mi

Answer: $\dfrac{20}{3}\pi$ mi$^2 \approx 20.9$ mi^2. ■

Accent on Technology: **Filling a water tank**

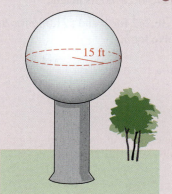

15 ft

FIGURE 143

See Figure 143. To calculate how many cubic feet of water are needed to fill a spherical water tank with a radius of 15 feet, we substitute 15 for r in the formula for the volume of a sphere.

$$V = \frac{4}{3}\pi r^3$$

$$V = \frac{4}{3}\pi(15)^3$$

To do the arithmetic with a graphing calculator, we enter these numbers and press these keys.

Keystrokes   4 ⊠ [2nd] [π] ⊠ 15 [^] 3 ÷ 3 [ENTER]   ⌈ 14137.16694 ⌉

To the nearest tenth, 14,137.2 ft^3 of water will be needed to fill the tank.

EXAMPLE 9   *Finding the diameter of a sphere.*   A sphere has a volume of 36π in.^{3}. What is its diameter?

Solution   To find the diameter of the sphere, we will find its radius and then double it.

$$V = \frac{4}{3}\pi r^3 \qquad \text{The formula for the volume of a sphere.}$$

$$36\pi = \frac{4}{3}\pi r^3 \qquad \text{Substitute } 36\pi \text{ for } V.$$

$$3 \cdot 36\pi = 3 \cdot \frac{4}{3}\pi r^3 \qquad \text{To clear the equation of the fraction, multiply both sides by 3.}$$

$$108\pi = 4\pi r^3 \qquad \text{Multiply: } 3 \cdot \frac{4}{3} = 4.$$

$$\frac{108\pi}{4\pi} = \frac{4\pi r^3}{4\pi} \qquad \text{To undo the multiplication by } 4\pi, \text{ divide both sides by } 4\pi.$$

$$27 = r^3 \qquad \text{Simplify: } \frac{108\pi}{4\pi} = \frac{27 \cdot \overset{1}{\cancel{4}} \cdot \overset{1}{\cancel{\pi}}}{\underset{1}{\cancel{4}} \cdot \underset{1}{\cancel{\pi}}} = 27.$$

$$r = 3 \qquad \text{What number cubed is 27? The answer is 3.}$$

Since the radius of the sphere is 3 inches, its diameter is $2 \cdot 3$ inches $= 6$ inches.   ■

Accent on Technology: **Volume of a silo**

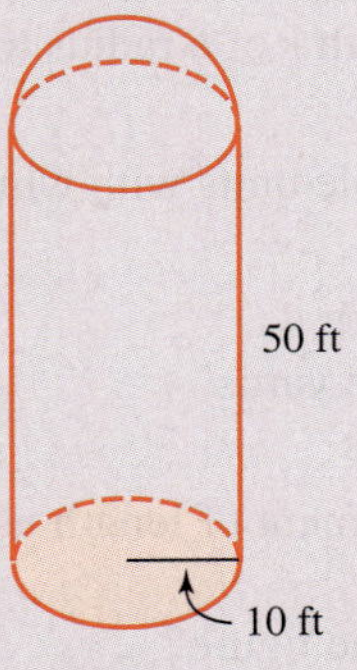

50 ft

10 ft

FIGURE 144

A silo is a structure used for storing grain. The silo in Figure 144 is a cylinder 50 feet tall topped with a **hemisphere** (a half-sphere). To find the volume of the silo, we add the volume of the cylinder to the volume of the dome.

$$\textbf{Volume}_{\textbf{cylinder}} + \textbf{Volume}_{\textbf{dome}} = (\textbf{Area}_{\textbf{cylinder's base}})(\textbf{Height}_{\textbf{cylinder}}) + \frac{1}{2}(\textbf{Volume}_{\textbf{sphere}})$$

$$= \pi r^2 h + \frac{1}{2}\left(\frac{4}{3}\pi r^3\right)$$

$$= \pi r^2 h + \frac{2\pi r^3}{3} \qquad \text{Multiply: } \frac{1}{2}\left(\frac{4}{3}\pi r^3\right) = \frac{4}{6}\pi r^3 = \frac{2\pi r^3}{3}.$$

$$= \pi(10)^2(50) + \frac{2\pi(10)^3}{3} \qquad \text{Substitute 10 for } r \text{ and 50 for } h.$$

To do the arithmetic with a graphing calculator, we enter these numbers and press these keys.

Keystrokes | 2nd | | π | × 10 | ^ | 2 × 50 | + | 2 × | 2nd | | π | × 10 | ^ | 3 | ÷ | 3

| ENTER | | 17802.35837 |

The volume of the silo is approximately 17,802 ft^3.

STUDY SET Section 10

VOCABULARY *Fill in the blanks.*

1. In _____, space is defined to be the set of all points.

2. Space figures are geometric figures that contain points in more than one _____.

3. If the lateral edges of a prism are perpendicular to its bases, it is called a _____ prism. Prisms that are not right prisms are called _____.

4. A right rectangular prism is referred to as a _____ solid.

5. If the bases and lateral faces of a prism are squares, the prism is called a _____.

6. A _____ is the set of all points in space that are a given distance, called the radius, from a given point, called the _____.

7. A _____ is one-half of a sphere.

8. If a polygon has sides that are all the same length and angles that have the same measure, we call it a _____ polygon.

9. The height of a pyramid or cone is the _____ distance from the vertex to the plane of the base.

10. The _____ of a three-dimensional figure is a measure of its capacity.

CONCEPTS

11. Give the complete name of each figure.

    a.

    b.

    c.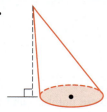

    d.

    e.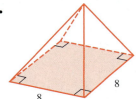

    f.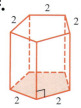

g. Copy Illustration 1. Then label it completely using the following words:

   base   base edge   lateral edge
   height   vertex   lateral face

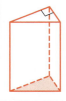

ILLUSTRATION 1

12. Draw a right hexagonal prism. Label the height *h*, the bases, and a lateral face.

13. Draw a cube. Label each base.

14. Draw a right circular cylinder. Label the height *h* and a radius *r*.

15. Draw an oblique square pyramid. Label the height *h*, the vertex, and the base.

16. Draw a right regular pentagonal pyramid. Label the height *h*, the vertex, and the base.

17. Draw an oblique circular cone. Label the height *h*, the vertex, and radius *r*.

18. Draw a sphere. Label a radius *r*.

19. Write a formula that relates the length *r* of a radius to the length *D* of a diameter of a circle.

20. Which of the following are acceptable units with which to measure volume?

ft^2	mi^3	seconds
cubic inches	mm	square yards
pounds	cm^2	meters

21. In Illustration 2, the unit of measurement of length used to draw the figure is the inch.
    a. What is the area of the base of the figure?
    b. What is the volume of the figure?

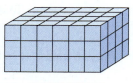

ILLUSTRATION 2

22. Which geometric concept (perimeter, circumference, area, or volume) should be applied when measuring each of the following?
    a. The distance around a checkerboard
    b. The size of a trunk of a car
    c. The amount of paper used for a postage stamp
    d. The amount of storage in a cedar chest
    e. The amount of beach available for sunbathing
    f. The distance the tip of a propeller travels

23. How many cubic inches are in 1 cubic foot?

24. How many cubic feet are in one 1 cubic yard?

25. How many cubic centimeters are in one cubic meter?

26. Complete the table.

Figure	Volume formula(s)
Cube	
Rectangular solid	
Prism	
Circular cylinder	
Pyramid	
Circular cone	
Sphere	

27. Evaluate each expression.
  a. $\frac{1}{3}(28)6$ **b.** $\frac{4}{3}(125)$

28. a. Evaluate $\frac{1}{3}\pi r^2 h$ for $r = 5$ and $h = 27$. Express your result in terms of π.
  b. Approximate your answer to part a to the nearest tenth.

29. If the height of a pyramid is doubled, how does its volume change?

30. If the radius of a cylinder is doubled, how does its volume change?

31. If the radius and the height of a cylinder are doubled, how does its volume change?

32. If the radius of a sphere is doubled, how does its volume change?

NOTATION

33. a. What does in.³ mean?
  b. Write "one cubic centimeter" using symbols.

34. In the formula $V = \frac{1}{3}Bh$, what does B represent?

35. In a drawing, what does the symbol $\llcorner$ indicate?

36. Redraw the figure in Illustration 3 using dashed lines to show the hidden edges.

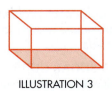

ILLUSTRATION 3

PRACTICE   *Find the volume of each figure. Give the exact answer and an approximate answer to the nearest hundredth, when applicable.*

37.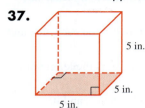

5 in.
5 in.
5 in.

38.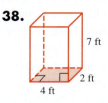

7 ft
2 ft
4 ft

39.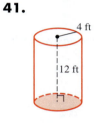

5 cm
6 cm
4 cm

40.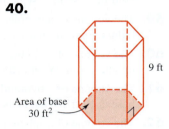

9 ft
Area of base 30 ft²

41.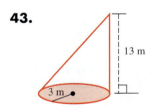

4 ft
12 ft

42.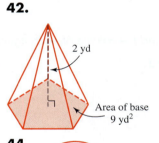

2 yd
Area of base 9 yd²

43.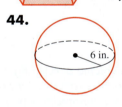

13 m
3 m

44.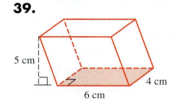

6 in.

Find the volume of each figure. Give the exact answer and an approximate answer to the nearest hundredth, when applicable.

45. A rectangular solid with dimensions of 3 cm by 4 cm by 5 cm.

46. A rectangular solid with dimensions of 5 m by 8 m by 10 m.

47. A prism whose base is a right triangle with legs 3 meters and 4 meters long and whose height is 8 meters.

48. A prism whose base is a right triangle with legs 5 feet and 12 feet long and whose height is 10 feet.

49. A sphere with a radius of 9 inches.

50. A sphere with a diameter of 10 feet.

51. A cylinder with a height of 12 meters and a circular base with a radius of 6 meters.

52. A cylinder with a height of 4 meters and a circular base with a diameter of 18 meters.

53. A cone with a height of 12 centimeters and a circular base with a diameter of 10 centimeters.

54. A cone with a height of 3 inches and a circular base with a radius of 4 inches.

55. A pyramid with a square base 10 meters on each side and a height of 12 meters.

56. A pyramid with a square base 6 inches on each side and a height of 4 inches.

57. The volume of a cube is 27 ft^3. What is the length of a side of the cube?

58. The area of the base of a prism is 16 in.^{2}, and its volume is 88 in.^{3}. What is the height of the prism?

59. The volume of a circular cylinder is 200π yd^3, and the radius of its circular base is 5 yd. What is its height?

60. The volume of a circular cylinder is 108π m^3, and its height is 9 m. What is the radius of its circular base?

61. The volume of a pyramid is 95 ft^3, and the area of its base is 19 ft^2. What is its height?

62. The volume of a sphere is 288π ft^3. Find its radius.

Find the volume of each figure. Express your answer in ft^3.

63.

60 in.

2 ft

1 yd

64.

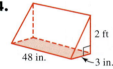

2 ft

48 in.

3 in.

Find the volume of each figure. Express your answer in m^3. Give the exact answer and an approximate answer to the nearest hundredth, when applicable.

65.

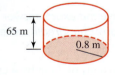

65 m

0.8 m

66.

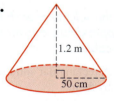

1.2 m

50 cm

Find the volume of each figure. Give the exact answer and an approximate answer to the nearest hundredth, when applicable.

67. Right regular hexagonal prism

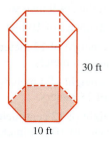

30 ft

10 ft

68. Right regular triangular pyramid

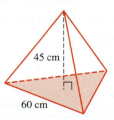

45 cm

60 cm

Find the volume of each figure. Give the exact answer and an approximate answer to the nearest hundredth, when applicable.

69.

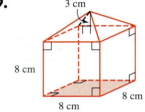

3 cm

8 cm

8 cm

8 cm

70.

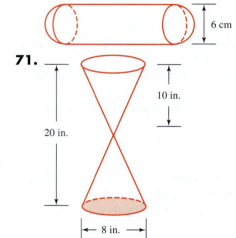

16 cm

6 cm

71.

20 in.

10 in.

8 in.

72.

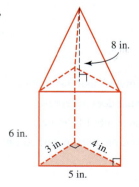

8 in.

6 in.

3 in.   4 in.

5 in.

APPLICATIONS   *Solve each problem. Give the exact answer and an approximate answer to the nearest hundredth, when applicable.*

73. VOLUME OF A SUGAR CUBE   A sugar cube is $\frac{1}{2}$ inch on each edge. How much volume does it occupy?

74. VOLUME OF A CLASSROOM   A classroom is 40 feet long, 30 feet wide, and 9 feet high. Find the number of cubic feet of air in the room.

75. WATER HEATER   Complete the advertisement for the high-efficiency water heater shown in Illustration 4.

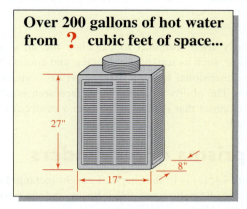

Over 200 gallons of hot water from ? cubic feet of space...

27"

17"   8"

ILLUSTRATION 4

76. REFRIGERATOR CAPACITY   The largest refrigerator advertised in a J. C. Penny catalog has a capacity of 25.2 cubic feet. How many cubic inches is this?

77. VOLUME OF AN OIL TANK   A cylindrical oil tank has a diameter of 6 feet and a length of 7 feet. Find the volume of the tank.

78. VOLUME OF A DESSERT   A restaurant serves pudding in a conical dish that has a diameter of 3 inches. If the dish is 4 inches deep, how many cubic inches of pudding are in each dish?

79. HOT-AIR BALLOON   The lifting power of a spherical balloon depends on its volume. How many cubic feet of gas will a balloon hold if it is 40 feet in diameter?

80. VOLUME OF A CEREAL BOX   A box of cereal measures 3 inches by 8 inches by 10 inches. The manufacturer plans to market a smaller box that measures $2\frac{1}{2}$ inches by 7 inches by 8 inches. By how much will the volume be reduced?

81. ENGINE   The *compression ratio* of an engine is the volume in one cylinder with the piston at bottom-dead-center (B.D.C.), divided by the volume with the piston at top-dead-center (T.D.C.). From the data given in Illustration 5, what is the compression ratio of the engine? Use a colon to express your answer.

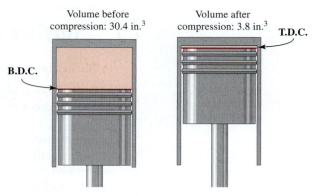

Volume before compression: 30.4 in.^{3} Volume after compression: 3.8 in.^{3} T.D.C.

B.D.C.

ILLUSTRATION 5

82. ESTIMATING THE VOLUME OF EARTH   Earth is not a perfect sphere but is slightly pear-shaped. To estimate its volume, we will assume that it is spherical, with a diameter of about 7,926 miles. What is its volume, to the nearest billion cubic miles?

WRITING

83. What is meant by the *volume* of a cube?

84. The stack of 3×5 index cards in Illustration 6 (a) forms a right rectangular prism, with a certain volume. If the stack is pushed to lean to the right, as in Illustration 6 (b), a new prism is formed. How will its volume compare to the volume of the right rectangular prism? Explain your answer.

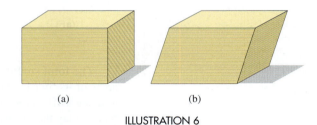

(a)   (b)

ILLUSTRATION 6

85. Are the units used to measure area different from the units used to measure volume? Explain.

86. The dimensions (length, width, and height) of one rectangular solid are entirely different numbers from the dimensions of another rectangular solid. Would it be possible for the rectangular solids to have the same volume? Explain.

11   *Surface Area*

In this section, you will learn about

- Surface areas of prisms and cylinders   • Surface areas of pyramids and cones
- Surface areas of spheres

INTRODUCTION. We have previously used formulas to calculate the areas of two-dimensional figures that lie in a plane, such as squares, rectangles, and circles. Now we will extend this concept to three-dimensional figures, such as prisms, cylinders, and spheres, to find their *surface area*. The ability to compute surface area is necessary when determining the amount of material that is needed to make a cardboard box, an aluminum can, or a plastic beach ball.

Surface areas of prisms and cylinders

The cardboard box shown in Figure 145(a) is in the shape of a right rectangular prism (rectangular solid). Recall that the top and bottom are called **bases,** and the front, back, left end, and right end are called **lateral faces** of the prism.

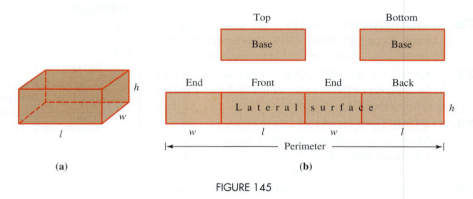

FIGURE 145

The **total surface area** *TSA* of the box is the sum of the area of its bases *and* the areas of its four lateral faces. It gives us a measure of the amount of cardboard needed to make it. To derive a formula for the surface area of such a figure, we disassemble the box and lay the pieces of cardboard out flat, as shown in Figure 145(b). We observe the following.

- The bases of the box are congruent. If *B* is the area of one base, then $B + B = 2B$ is the sum of the areas of both of its bases.
- The sum of the areas of the four lateral faces of the box is called its **lateral surface area** *LSA*. One way to find the *LSA* is to compute the area of each of the lateral faces and then add them. Or we can use the fact that, when laid out flat, the lateral faces form one large rectangle. (See Figure 145(b).) The width of the rectangle is the height of the box; the length of the rectangle is the perimeter of the base of the box. To find the lateral surface area, we can simply multiply the height *h* of the box by the perimeter *p* of its base: $LSA = hp$.

These observations suggest a formula to find the total surface area of any prism.

Surface area of a prism

> The **total surface area** *TSA* of a prism is the sum of the area of its bases and its lateral surface area. If a prism has height *h* and if each base has a perimeter *p* and area *B*, the formula for the total surface area is given by
>
> $TSA = 2B + hp$

COMMENT   As with any type of measurement of area, surface area is measured in square units, such as square feet (ft²), square inches (in.²), and square centimeters (cm²).

EXAMPLE 1   *Surface area of a triangular prism.*
Find the total surface area of the prism shown in Figure 146.

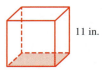

FIGURE 146

Solution

To find the total surface area of the prism, we need to know the area of one of its bases, the perimeter of a base, and the height of the figure.

To find the area of one of its triangular bases, we substitute 10 for b and 24 for h in the formula for the area of a triangle.

$$A = \frac{1}{2}bh$$

$$A = \frac{1}{2}(10)(24)$$ One leg of the right triangle is the base, and the other leg is the height of the triangle.

$$= 5(24)$$ Multiply: $\frac{1}{2}(10) = 5$.

$$= 120$$

The area of one base of the prism is 120 ft².

The perimeter p (in feet) of a base is

$$p = 10 + 24 + 26 = 60$$

To find the total surface area of the prism, we proceed as follows.

$TSA = 2B + hp$ The formula for the surface area of a prism.

$TSA = 2(120) + 15(60)$ Substitute 120 for B, 15 for h, and 60 for p.

$TSA = 240 + 900$ The area of the bases is 240 ft², and the *LSA* is 900 ft².

$TSA = 1,140$

The total surface area of the prism is 1,140 ft².

Self Check

Find the total surface area of the cube shown below.

11 in.

Answer: 726 in.²   ■

EXAMPLE 2   *Finding the area of the base of a prism.*   Find the area of one base of the right regular octagonal prism in Figure 147 if its total surface area is 200 m².

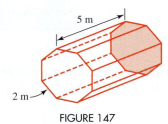

5 m

2 m

FIGURE 147

Solution   The prism is lying on one of its lateral faces. Since it is a regular octagon, each of the eight sides is 2 m long, and the perimeter of the base is 8 · 2 m = 16 m. We substitute 200 for *TSA*, 5 for h, and 16 for p in the formula for the total surface area of a prism and solve for B, the area of one base.

$$TSA = 2B + hp$$
$$200 = 2B + 5(16)$$
$$200 = 2B + 80$$
$$120 = 2B \qquad \text{Subtract 80 from both sides.}$$
$$60 = B \qquad \text{Divide both sides by 2.}$$

The area of one base of the prism is 60 m^2.   ∎

To determine the amount of material need to make the aluminum can in Figure 148(a), we need to find its surface area—the sum of the areas of its bases and its lateral surface area.

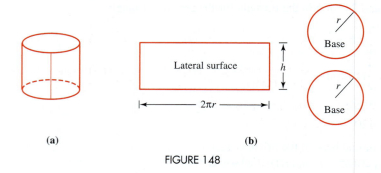

(a)   (b)

FIGURE 148

To derive a formula for the surface area of a right circular cylinder, we will take an approach similar to the one we used with prisms. In Figure 148(b), the can is disassembled, and the pieces of aluminum are laid out flat. We observe the following.

- The area of one of the circular bases is πr^2. Therefore, the sum of the areas of two circular bases is $\pi r^2 + \pi r^2 = 2\pi r^2$.
- When the lateral surface of the can is "unrolled," its shape is rectangular. The area of the rectangle is the product of its length $2\pi r$ (the circumference of one circular base) and the height h of the can: $LSA = 2\pi rh$.

These observations suggest the following formula.

Surface area of a right circular cylinder

The **total surface area** *TSA* of a right circular cylinder is the sum of the area of its two circular bases and its lateral surface area. If a right circular cylinder has height h and if the radius of the base is r, then the total surface area is given by

$$TSA = 2\pi r^2 + 2\pi rh$$

EXAMPLE 3 *Total surface area of a cylinder.* Find the total surface area of the right circular cylinder in Figure 149.

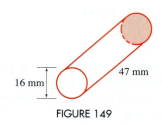

16 mm   47 mm

FIGURE 149

Self Check

To the nearest tenth, find the total surface area of the right circular cylinder shown below.

3 cm

7.5 cm

Solution

In this position, one of the cylinder's circular bases is facing the reader. Since its diameter is 16 millimeters, the radius r is 8 mm. If the cylinder were standing vertically, its height h would be 47 mm. To find the total surface area, we proceed as follows.

$$TSA = 2\pi r^2 + 2\pi rh$$
$$TSA = 2\pi(8)^2 + 2\pi(8)(47)$$
$$= 2\pi(64) + 2\pi(376)$$
$$= 128\pi + 752\pi \qquad \text{The area of the bases is } 128\pi \text{ mm}^2, \text{ and the } LSA \text{ is}$$
$$\qquad\qquad\qquad\qquad 752\pi \text{ mm}^2.$$
$$= 880\pi$$

The total surface area is 880π mm². To the nearest tenth, this is 2,764.6 mm².

Answer: $63\pi \text{ cm}^2 \approx 197.9 \text{ cm}^2$ ∎

Surface areas of pyramids and cones

Recall that a *right regular pyramid* is a pyramid whose base is a regular polygon and whose vertex is equidistant from each vertex of the base. Figure 150(a) shows a regular pyramid—more specifically, a right regular square pyramid. The four lateral faces of the pyramid are congruent triangles. The height s of each lateral face is called the **slant height** of the pyramid.

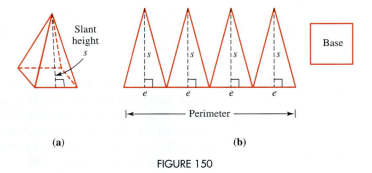

(a)   **(b)**

FIGURE 150

To derive a formula for the total surface area of a right regular pyramid, we can "unfold" the figure and lay the pieces out flat, as shown in Figure 150(b). If the length of each base edge is e, then the area of each triangle is one-half of the product of the length of its base and the slant height, or $\frac{1}{2}es$. To find the lateral surface area of the pyramid, we need to find the sum of areas of the four triangular lateral faces.

$$LSA = \tfrac{1}{2}es + \tfrac{1}{2}es + \tfrac{1}{2}es + \tfrac{1}{2}es$$

$$LSA = \tfrac{1}{2}s(e + e + e + e) \qquad \text{Factor out the common factor of } \tfrac{1}{2}s.$$

The expression $e + e + e + e$ is simply the sum of the lengths of the four edges of the base. We can replace it with the variable p, where p is the perimeter of the base.

$$LSA = \tfrac{1}{2}sp \qquad \text{Because } e + e + e + e = p.$$

$$LSA = \tfrac{1}{2}ps \qquad \text{Apply the commutative property of multiplication to write the variable factors in alphabetical order.}$$

To find the total surface area of the pyramid, we add the area of the base to the lateral surface area.

Surface area of a pyramid

> The **total surface area** *TSA* of a right regular pyramid is the sum of the area of its base and its lateral surface area. If *B* is the area of the base of a right regular pyramid, *p* the perimeter of the base, and *s* the slant height, then the total surface area is given by
>
> $$TSA = B + \frac{1}{2}ps$$

EXAMPLE 4   *Surface area of a pyramid.*   Find the surface area of the right regular pyramid shown in Figure 151(a), if its base is an equilateral triangle with sides 8 inches long and its slant height is 11 inches.

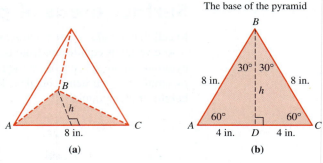

FIGURE 151

Solution   Figure 151(b) shows the base of the pyramid, which we will call $\triangle ABC$. Because $\triangle ABC$ is equilateral, each side is 8 inches long, and each angle has measure 60°. To find its height *h,* we construct altitude $\overline{BD}$ that divides the equilateral triangle into two congruent 30°–60°–90° triangles. Since $\overline{BD}$ is the longer leg and $\overline{DC}$ is the shorter leg of $\triangle DBC$, and since $m(\overline{DC}) = 4$ inches, it follows that $h = 4\sqrt{3}$ inches. Knowing *h,* we can now find the area of $\triangle ABC$.

$$\text{Area of } \triangle ABC = \frac{1}{2}bh$$

$$= \frac{1}{2}(8)(4\sqrt{3})$$

$$= 16\sqrt{3}$$

The area of the base of the pyramid is $16\sqrt{3}$ in.^{2}.

To find the surface area of the pyramid, we note that the area of the base is $16\sqrt{3}$ in.^{2}, the perimeter of the base is 8 in. + 8 in. + 8 in. = 24 in., and the slant height is 11 in.

$$TSA = B + \frac{1}{2}ps \qquad \text{\color{red}The formula for the surface area of a pyramid.}$$

$$TSA = 16\sqrt{3} + \frac{1}{2}(24)(11) \quad \text{\color{red}Substitute } 16\sqrt{3} \text{ for } B, 24 \text{ for } p, \text{ and } 11 \text{ for } s.$$

$$= 16\sqrt{3} + 12(11)$$

$$= 16\sqrt{3} + 132 \qquad \text{\color{red}The area of the base is } 16\sqrt{3} \text{ in.}^2, \text{ and the } LSA \text{ is } 132 \text{ in.}^2.$$

The total surface area of the pyramid is exactly $(16\sqrt{3} + 132)$ in.^{2}. To the nearest tenth, this is 159.7 in.^{2}.

Figure 152(a) shows a right circular cone. Right circular cones have both a height and a **slant height.** The slant height *s* is the length of a line segment that joins the vertex of the cone to a point on the edge of its circular base.

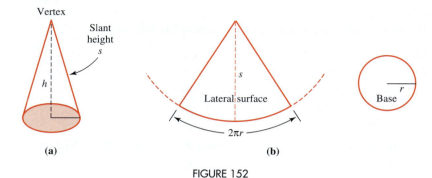

FIGURE 152

To derive the formula for the surface area of a right circular cone, we disassemble it and lay the pieces out flat, as shown in Figure 152(b). We observe the following.

- The area of its circular base is πr^2.

- When the lateral surface of the can is "unrolled," a figure called a *sector* results. This sector is almost triangular-shaped. The length of the base of this "triangle" is the circumference of the circular base of the cone. The height of this "triangle" is the slant height *s* of the cone. Therefore, the area of the sector is $\frac{1}{2}bh = \frac{1}{2}(2\pi r)s = \pi rs$.

These observations suggest the following formula.

Surface area of a cone

> The **total surface area** *TSA* of a right circular cone is the sum of the area of its base and its lateral surface area. If *r* is the radius of the base and *s* the slant height, then the total surface area of the cone is given by
>
> $$TSA = \pi r^2 + \pi rs$$

EXAMPLE 5   *Surface area of a cone.* Find the surface area of the right circular cone shown in Figure 153.

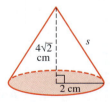

FIGURE 153

Self Check

To the nearest tenth, find the surface area of the right circular cone shown below.

Solution

From the figure, we see that the segments representing the height, a radius, and a slant height form a right triangle. We can use the Pythagorean theorem to find the unknown slant height *s* of the cone.

$$a^2 + b^2 = c^2$$
$$2^2 + \left(4\sqrt{2}\right)^2 = s^2 \quad \text{Substitute 2 for } a, 4\sqrt{2} \text{ for } b, \text{ and } s \text{ for } c.$$
$$4 + 32 = s^2 \quad \left(4\sqrt{2}\right)^2 = \left(4\sqrt{2}\right)\left(4\sqrt{2}\right) = 16 \cdot 2 = 32.$$
$$36 = s^2$$

By the square root property, $s = \sqrt{36} = 6$ or $s = -\sqrt{36} = -6$. Since the slant height must be positive, we have

$$s = 6$$

The slant height of the cone is 6 cm.

To find the surface area of the cone, we proceed as follows.

$TSA = \pi r^2 + \pi rs$	The formula for the surface area of a right circular cone.
$TSA = \pi(2)^2 + \pi(2)(6)$	The radius r of the base is 2 cm, and the slant height s is 6.
$= 4\pi + 12\pi$	The area of the base of the cone is 4π cm^2, and its *LSA* is 12π cm^2.
$= 16\pi$	

The total surface of the cone is 16π cm^2. To one decimal place, this is 50.3 cm^2.

Answer: $96\pi\,\text{m}^2 \approx 301.6\,\text{m}^2$ ■

Surface areas of spheres

We can think of a sphere as a hollow ball. More formally, a sphere is the set of all points that lie a fixed distance r from a point called the *center*. A segment drawn from the center of the sphere to a point on the sphere is called a *radius*. There is a formula to find the surface area of a sphere.

Surface area of a sphere

> The **surface area** *SA* of a sphere with radius r is given by
>
> $$SA = 4\pi r^2$$

EXAMPLE 6   *Manufacturing beach balls.*   A beach ball is to have a diameter of 16 inches. (See Figure 154.) How many square inches of material will be needed to make the ball? (Disregard any waste.)

Solution

Since a radius r of the ball is one-half the diameter, $r = 8$ inches. We can now substitute 8 for r in the formula for the surface area of a sphere.

$$SA = 4\pi r^2$$
$$SA = 4\pi(8)^2$$
$$SA = 4\pi(64)$$

$SA = 256\pi$	Multiply: $4 \cdot 64 = 256$.
≈ 804.2477193	Use a calculator.

A little more than 804 in.^{2} of material is needed to make the ball.

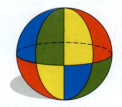

FIGURE 154

Self Check

Find the surface area of a beach ball with a diameter twice that of the ball shown in Figure 154.

Answer: $1{,}024\pi\,\text{in.}^2 \approx 3{,}217\,\text{in.}^2$

■

STUDY SET   Section 11

VOCABULARY   *Fill in the blanks.*

1. The total _____ _____ of a prism is the sum of the area of its bases and its lateral faces.

2. Copy the figure in Illustration 1. Label the vertex, the slant height s, and the height h of the pyramid.

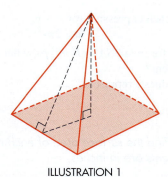

ILLUSTRATION 1

3. The _____ surface area of a pyramid is the sum of the area of its triangular lateral faces.

4. A right _____ pyramid is a pyramid whose base is a regular polygon and whose vertex is equidistant from each vertex of the base.

5. The total surface area of a right circular cylinder is the sum of the area of its two circular _____ and its lateral surface area.

6. Copy the figure in Illustration 2. Label the vertex, the slant height s, the height h of the cone, and the radius r of the base of the cone.

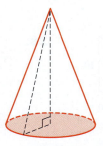

ILLUSTRATION 2

7. If a segment from the vertex of a cone to the center of the base of the cone is perpendicular to the base, the cone is called a _____ circular cone.

8. A _____ is the set of all points that lie a fixed distance r from a point called the center.

CONCEPTS

9. Which of the following are acceptable units of measurement for surface area?

ft^2	mi^3	seconds
cubic inches	mm	square yards
gallons	cm^2	meters

10. Suppose the area of the base of a pyramid is 55 ft^2 and its lateral surface area is 144 ft^2. What is the total surface area of the figure?

11. Illustration 3 shows a right rectangular prism that has been "disassembled" and laid out flat. Fill in the blanks.

a. When we find the area of the blue-shaded regions, we are finding the area of the _____.

b. When we find the area of the red-shaded regions, we are finding the _____ surface area.

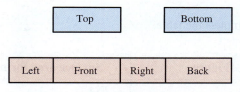

ILLUSTRATION 3

12. Draw a picture of each figure after it has been "disassembled" and laid out flat. Label the bases and the lateral surface area.

a. Cube   **b.** Right circular cone

c. Right circular cylinder   **d.** Square pyramid

 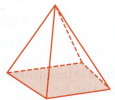

13. What is the perimeter p of the base of the right regular hexagonal prism in Illustration 4?

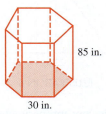

85 in.

30 in.

ILLUSTRATION 4

14. Give the formula that can be used to find the area B of the base of each figure.

a. Right square prism **b.** Regular triangular pyramid

 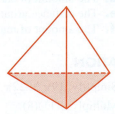

c. Right rectangular prism   **d.** Right circular cone

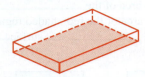

15. Complete the table.

Figure	Surface area formula
Prism	$TSA =$
Right circular cylinder	$TSA =$
Regular pyramid	$TSA =$
Right circular cone	$TSA =$
Sphere	$TSA =$

16. In Illustration 5, find *r*.

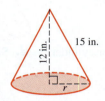

ILLUSTRATION 5

17. Refer to the equilateral triangle in Illustration 6.
   a. Find *h*.
   b. To the nearest hundredth, find the area of the triangle.

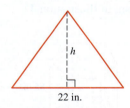

22 in.

ILLUSTRATION 6

18. Which geometric concept (perimeter, circumference, area, volume, or surface area) should be applied to find each of the following?
   a. The size of a room to be air conditioned
   b. The amount of land in a national park
   c. The amount of space in a refrigerator freezer
   d. The amount of cardboard in a shoe box
   e. The distance around a checkerboard
   f. The amount of material used to make a basketball

NOTATION

19. Simplify: $19\pi + 22\pi$.

20. Multiply: $\frac{1}{2}(11)(8)$.

21. Evaluate each expression for $r = 6$.
   a. $4\pi r^2$ **b.** $2\pi r^2$

22. a. In the formula $TSA = B + \frac{1}{2}ps$, what does B represent?
   b. What does s represent?

PRACTICE   *Find the surface area of each right prism. The measurements are in inches.*

23.

8
2
2

24.

9
7
15

25.

36   15
50
39

26.
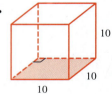
26
5
13
12

27.
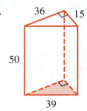
10
10
10

28.
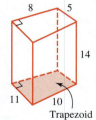
8   5
14
11   10
Trapezoid

The area of one base of a right regular hexagonal prism is given. Find its total surface area. Give the exact answer and an approximate answer. Round to the nearest tenth.

29.
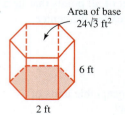
Area of base
$24\sqrt{3}$ ft²
6 ft
2 ft

30.
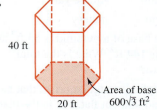
40 ft
Area of base
$600\sqrt{3}$ ft²
20 ft

The base of each right prism is an equilateral triangle. Find the area of one base of the prism. Then find its total surface area. Give the exact answer and an approximate answer. Round to the nearest tenth. The measurements are in centimeters.

31.
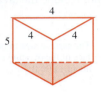

32.

Find the surface area of each right circular cylinder. Give the exact answer and an approximate answer. Round to the nearest tenth. The measurements are in feet.

33.

34.

35.

36.
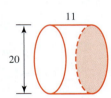

Find the surface area of each regular square pyramid. The measurements are in yards.

37.
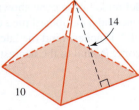

38.
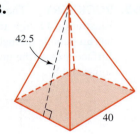

39.
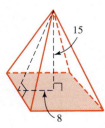

40.
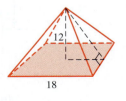

Find the area of the base of the regular pyramid. Then find its total surface area. Give the exact answer and an approximate answer. Round to the nearest tenth. The measurements are in meters.

41.
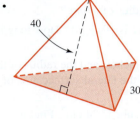

42.
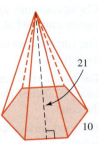

Find the surface area of each right circular cone. Give the exact answer and an approximate answer. Round to the nearest tenth. The measurements are in inches.

43.

44.
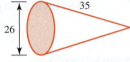

45.
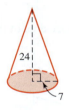

46.
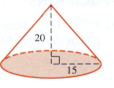

Find the surface area of each sphere. Give the exact answer and an approximate answer. Round to the nearest tenth. The measurements are in inches.

47.

48.

49.

50.

51. The total surface area of a right square prism is 230 ft^2. Find the height of the prism if the length of a side of a base is 5 ft.

52. The total surface area of a right circular cylinder is 96π in.^{2}. If the radius of its base is 6 in., what is its height?

53. The lateral surface area of a regular square pyramid is 240 m^2. If each side of its base is 10 m long, what is its slant height?

54. The area of the base of a right circular cone is 400π cm^2. If the slant height is 29 cm, find the height of the cone.

55. If the lateral surface area of a right circular cylinder is 42π square units and the total surface area is 60π square units, what is the radius of a base?

56. The surface area of a sphere is 196π yd^2. How long is its radius?

57. The height of a cone is 24 inches, and the radius of the base is 7 inches long. Find the lateral surface area of the cone.

58. The total surface area of a cube is 24 cm^2. Find its volume.

59. Find the total surface area of a right prism whose bases are regular hexagons with sides 12 millimeters long and whose height is 10 mm.

60. The base of a right prism is a rhombus with diagonals measuring 10 feet and 24 feet. The height of the prism is 60 feet. Find the total surface area of the figure.

APPLICATIONS

61. PENCILS   Determine the lateral surface area of the pencil shown in Illustration 7.

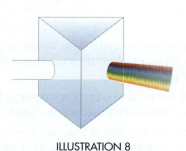

170 mm

4 mm

ILLUSTRATION 7

62. LIGHT   A triangular prism separates white light into a visible spectrum composed of primary colors. In Illustration 8, the height of the prism is 10 mm, and each edge of its base is 2 mm long. Find the total surface area. Round to the nearest tenth.

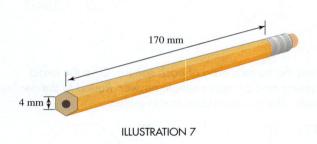

ILLUSTRATION 8

63. WASHINGTON, D.C.   The Washington Monument is a tall shaft of marble blocks that is topped by a regular square pyramid. To find its surface area, we can "disassemble" it as shown in Illustration 9. Use the given information to estimate the lateral surface area of the monument. (In this case, do not include the area of its base.)

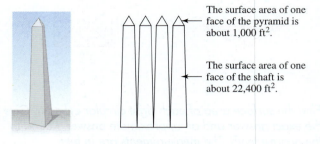

The surface area of one face of the pyramid is about 1,000 ft^2.

The surface area of one face of the shaft is about 22,400 ft^2.

ILLUSTRATION 9

64. LINT REMOVER   Illustration 10 shows a handy gadget; it uses a cylinder of sheets of sticky paper that can be rolled over clothing and furniture to pick up lint and pet hair. After the paper is full, that sheet is peeled away to expose another sheet of sticky paper. Find the area of the first sheet.

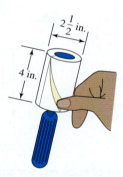

$2\frac{1}{2}$ in.

4 in.

ILLUSTRATION 10

65. ROCKETRY   The model rocket nose cone shown in Illustration 11 slips into one end a cardboard tube that has an outside diameter of 2.4 inches. Find the lateral surface area of the nose cone. Round to the nearest square inch.

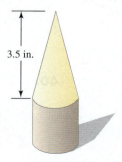

3.5 in.

ILLUSTRATION 11

66. BOWLING   A bowling ball is packaged within a tightly fitting cubical cardboard box, as shown in Illustration 12. Approximately how many times greater is the surface area of the box compared to the surface area of the ball?

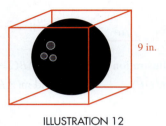

9 in.

ILLUSTRATION 12

WRITING

67. What is meant by *lateral surface area* of a figure? Give an example.

68. Explain how the Pythagorean theorem is used in this section.

69. Explain the difference between the *surface area* and the *volume* of a figure.

70. A right rectangular prism has length 2 ft, width 3 ft, and height 4 ft. Suppose each dimension is then doubled. How much greater will the surface area of the new prism be compared to that of the original prism? Explain how you arrive at your answer.

Sections 8-11 *Cumulative Review Exercises*

In Exercises 1–10, match each concept in Column I with the most appropriate response in Column II.

1. Pythagorean theorem

a. $\dfrac{a}{b} = \dfrac{c}{d}$

2. Congruent triangles

b. Corresponding sides are proportional.

3. Volume

c. 45°–45°–90° triangle

4. Proportion

d.

5. Right isosceles triangle

e. Six pairs of congruent corresponding parts

6. 30°–60°–90° triangle

f. Lateral surface area

7. Similar triangles

g. Capacity

8. Prism

h. $a^2 + b^2 = c^2$

9. LSA

i. Shorter leg half as long as hypotenuse

10. Slant height

j.

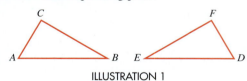

11. See Illustration 1, where $\triangle ABC \cong \triangle DEF$. Complete the list of corresponding parts.

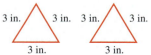

ILLUSTRATION 1

∠A corresponds to _____.

∠B corresponds to _____.

∠C corresponds to _____.

$\overline{AC}$ corresponds to _____.

$\overline{AB}$ corresponds to _____.

$\overline{BC}$ corresponds to _____.

12. Tell whether the triangles in each pair are congruent. If they are, tell why.

a.

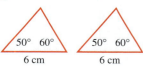

b.

3 in. 3 in. 3 in. 3 in.

3 in. 3 in.

50° 60°   50° 60°

6 cm   6 cm

c.

70°
60° 50°

70°
60° 50°

d.

70°

70°

13. Refer to Illustration 2, in which $\triangle ABC \cong \triangle DEF$.

a. Find m($\overline{DE}$).

b. Find m($\angle E$).

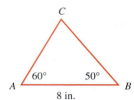

ILLUSTRATION 2

14. Tell whether the triangles in each pair are similar.

a.

50° 50°
50° 50°

b.

35°

35°

15. Refer to Illustration 3. Find x and y.

a. Find x.

b. Find y.

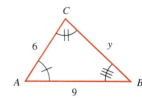

ILLUSTRATION 3

16. SHADOWS   If a tree casts a 7-foot shadow at the same time as a man 6 feet tall casts a 2-foot shadow, how tall is the tree?

17. Refer to Illustration 4. Find the length of the unknown side.

a. If $a = 10$ and $b = 24$, find c.

b. If $a = 6$ and $c = 8$, find b.

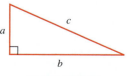

ILLUSTRATION 4

18. Find the missing lengths in each triangle. Give the exact answer and then an approximation to two decimal places when necessary.

a.

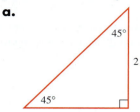

b.

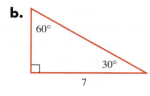

c.

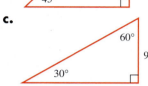

19. TELEVISION   To the nearest tenth of an inch, what is the diagonal measurement of the television screen in Illustration 5?

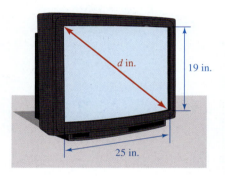

ILLUSTRATION 5

20. CAR REPAIR   To create some space to work under the front end of a car, a mechanic drives it up steel ramps. See Illustration 6. The ramp is 1 foot longer than the back, and the base is 2 feet longer than the back of the ramp. Find the length of each side of the ramp.

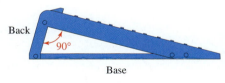

ILLUSTRATION 6

21. How many cubic inches are there in 1 cubic foot?

22. Find the volume of each figure. Give the exact answer and an approximate answer to the nearest hundredth, when applicable.

a.

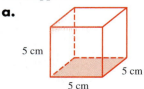

b.

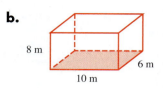

c.

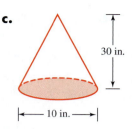

d.

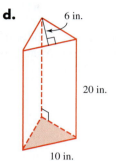

e.

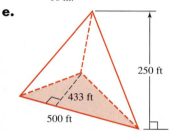

23. FARMING   A silo is used to store wheat and corn. Find the volume of the silo shown in Illustration 7. Round to the nearest unit.

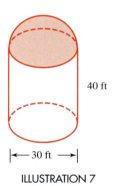

ILLUSTRATION 7

24. A square pyramid has a volume of 48 in.^{3}. Find its height if the length of a side of its base is 6 inches.

25. Find the surface area of each figure. Give the exact answer and an approximate answer to the nearest hundredth, when applicable.

a.

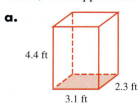

b.

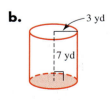

c.

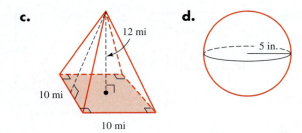

12 mi

10 mi

10 mi

d.

5 in.

27. Give a real-life example in which the concept of perimeter is used. Do the same for area and for volume. Be sure to discuss the type of units used in each case.

28. What is the difference between lateral surface area and total surface area of a pyramid? Make a drawing to help explain your answer.

26. The surface area of a right circular cone is 880π ft^2. The radius of its circular base is 20 feet. What is its slant height?

In this appendix, you will learn about

- Inductive reasoning • Deductive reasoning

INTRODUCTION. To reason means to think logically. The objective of this appendix is to develop your problem-solving ability by improving your reasoning skills. We will introduce two fundamental types of reasoning that can be applied in a wide variety of settings. They are known as *inductive reasoning* and *deductive reasoning.*

Inductive reasoning

In a laboratory, scientists conduct experiments and observe outcomes. After several repetitions with similar outcomes, the scientist will generalize the results into a statement that appears to be true:

- If I heat water to 212°F, it will boil.
- If I drop a weight, it will fall.
- If I combine an acid with a base, a chemical reaction occurs.

When we draw general conclusions from specific observations, we are using **inductive reasoning.** The next examples show how inductive reasoning can be used in mathematical thinking. Given a list of numbers or symbols, called a *sequence,* we can often find a missing term of the sequence by looking for patterns and applying inductive reasoning.

EXAMPLE 1 *An increasing pattern.* Find the next number in the sequence 5, 8, 11, 14, ...

Solution
The terms of the sequence are increasing. To discover the pattern, we find the *difference* between each pair of successive terms.

$8 - 5 = 3$ Subtract the first term from the second term.

$11 - 8 = 3$ Subtract the second term from the third term.

$14 - 11 = 3$ Subtract the third term from the fourth term.

The difference between each pair of numbers is 3. This means that each successive number is 3 greater than the previous one. Thus, the next number in the sequence is $14 + 3$, or 17.

Self Check
Find the next number in the sequence $-3, -1, 1, 3, ...$

Answer: 5

A-1

EXAMPLE 2 *A decreasing pattern.* Find the next number in the sequence $-2, -4, -6, -8, \ldots$

Solution

The terms of the sequence are decreasing. Since each successive term is 2 less than the previous one, the next number in the pattern is $-8 - 2$, or -10.

Self Check

Find the next number in the sequence $-0.1, -0.3, -0.5, -0.7 \ldots$

Answer: -0.9 ◼

EXAMPLE 3 *An alternating pattern.* Find the next letter in the sequence A, D, B, E, C, F, D, ...

Solution

The letter A is the first letter of the alphabet, D is the fourth letter, B is the second letter, and so on. We can create the following letter–number correspondence:

$$
\begin{array}{ll}
A \rightarrow 1 & \\
D \rightarrow 4 & \text{Add 3.} \\
B \rightarrow 2 & \text{Subtract 2.} \\
E \rightarrow 5 & \text{Add 3.} \\
C \rightarrow 3 & \text{Subtract 2.} \\
F \rightarrow 6 & \text{Add 3.} \\
D \rightarrow 4 & \text{Subtract 2.}
\end{array}
$$

The numbers in the sequence 1, 4, 2, 5, 3, 6, 4, . . . alternate in size. They change from smaller to larger, to smaller, to larger, and so on.

We see that 3 is added to the first number to get the second number. Then 2 is subtracted from the second number to get the third number. To get successive terms in the sequence, we alternately add 3 to one number and then subtract 2 from that result to get the next number.

Applying this pattern, the next number in the numerical sequence would be $4 + 3$, or 7. The next letter in the original sequence would be G, because it is the seventh letter of the alphabet.

Self Check

Find the next entry in the sequence Z, A, Y, B, X, C, ...

Answer: W   ◼

EXAMPLE 4 *Two patterns.* Find the next geometric shape in the sequence below.

Solution

This sequence has two patterns occurring at the same time. The first figure has three sides and one dot, the second figure has four sides and two dots, and the third figure has five sides and three dots. Thus, we would expect the next figure to have six sides and four dots, as shown in Figure A-1.

FIGURE A-1

Self Check

Find the next geometric shape in the sequence below.

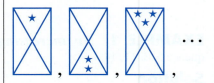

Answer: ◼

EXAMPLE 5   *A circular pattern.*   Find the next geometric shape in the sequence below.

Self Check

Find the next geometric shape in the sequence below.

Solution

From figure to figure, we see that each dot moves from one point of the star to the next, in a counterclockwise direction. This is a circular pattern. The next shape in the sequence will be the one shown in Figure A-2.

FIGURE A-2

Answer:

Deductive reasoning

As opposed to inductive reasoning, **deductive reasoning** moves from the general case to the specific. For example, if we know that the sum of the angles in any triangle is 180°, we know that the sum of the angles of △*ABC* is 180°. Whenever we apply a general principle to a particular instance, we are using deductive reasoning.

A deductive reasoning system is built on four elements:

1. **Undefined terms:** terms that we accept without giving them formal meaning
2. **Defined terms:** terms that we define in a formal way
3. **Axioms** or **postulates:** statements that we accept without proof
4. **Theorems:** statements that we can prove with formal reasoning

Many problems can be solved by deductive reasoning. For example, suppose that we plan to enroll in an early-morning algebra class, and that we know that Professors Perry, Miller, and Tveten are scheduled to teach algebra next semester. After some investigating, we find out that Professor Perry teaches only in the afternoon and Professor Tveten teaches only in the evenings. Without knowing anything about Professor Miller, we can conclude that he will be our teacher, since he is the only remaining possibility.

The following examples show how to use deductive reasoning to solve problems.

EXAMPLE 6   *Scheduling classes.*   Four professors are scheduled to teach mathematics next semester, with the following course preferences:

1. Professors A and B don't want to teach calculus.
2. Professor C wants to teach statistics.
3. Professor B wants to teach algebra.

Who will teach trigonometry?

Solution   The following chart shows each course, with each possible instructor.

Calculus	Algebra	Statistics	Trigonometry
A	A	A	A
B	B	B	B
C	C	C	C
D	D	D	D

Since Professors A and B don't want to teach calculus, we can cross them off the calculus list. Since Professor C wants to teach statistics, we can cross her off every other list. This leaves Professor D as the only person to teach calculus, so we can cross her off every other list. Since Professor B wants to teach algebra, we can cross him off every other list. Thus, the only remaining person left to teach trigonometry is Professor A.

Calculus	Algebra	Statistics	Trigonometry
A̶	A	A̶	A
B̶	B	B̶	B̶
C̶	C̶	C	C̶
D	D̶	D̶	D̶

**EXAMPLE 7   *State flags.*   The graph in Figure A-3 gives the number of state flags that feature an eagle, a star, or both. How many state flags have neither an eagle nor a star?

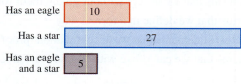

FIGURE A-3

Solution

In Figure A-4(a), the intersection (overlap) of the circles is a way to show that there are 5 state flags that have both an eagle and a star. If an eagle appears on a total of 10 flags, then the left circle must contain 5 more flags outside of the intersection. See Figure A-4(b). If a total of 27 flags have a star, the right circle must contain 22 more flags outside the intersection.

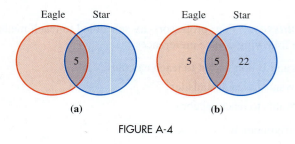

FIGURE A-4

Self Check

Of the 50 cars on a used-car lot, 9 are red, 31 are foreign models, and 6 are red, foreign models. If a customer wants to buy an American model that is not red, how many cars does she have to choose from?

From Figure A-4, we see that $5 + 5 + 22$, or 32 flags have an eagle, a star, or both. To find how many flags have neither an eagle nor a star, we subtract this total from the number of state flags, which is 50.

$$50 - 32 = 18$$

There are 18 state flags that have neither an eagle nor a star.

Answer: 16   ■

Study Set   Appendix I

VOCABULARY   *Fill in the blanks.*

1. _____ reasoning draws general conclusions from specific observations.

2. _____ reasoning moves from the general case to the specific.

CONCEPTS   *Tell whether the pattern shown is increasing, decreasing, alternating, or circular.*

3. $2, 3, 4, 2, 3, 4, 2, 3, 4, \ldots$

4. $8, 5, 2, -1, \ldots$

5. $-2, -4, 2, 0, 6, \ldots$

6. $0.1, 0.5, 0.9, 1.3, \ldots$

7. $a, c, b, d, c, e, \ldots$

8.  $\ldots$

9. ROOM SCHEDULING   From the chart, determine what time(s) on a Wednesday morning a practice room in a music building is available. The symbol X indicates that the room has already been reserved.

	M	T	W	Th	F
9 A.M.	X		X	X	
10 A.M.	X	X			X
11 A.M.			X		X

10. COUNSELING QUESTIONNAIRE   A group of college students were asked if they were taking a mathematics course and if they were taking an English course. The results are displayed in Illustration 1.

   a. How many students were taking a mathematics course and an English course?

   b. How many students were taking an English course but not a mathematics course?

   c. How many students were taking a mathematics course?

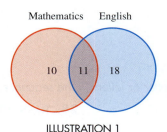

   Mathematics   English

   ILLUSTRATION 1

PRACTICE   *Find the number that comes next in each sequence.*

11. $1, 5, 9, 13, \ldots$

12. $15, 12, 9, 6, \ldots$

13. $-3, -5, -8, -12, \ldots$

14. $5, 9, 14, 20, \ldots$

15. $-7, 9, -6, 8, -5, 7, -4, \ldots$

16. $2, 5, 3, 6, 4, 7, 5, \ldots$

17. $9, 5, 7, 3, 5, 1, \ldots$

18. $1.3, 1.6, 1.4, 1.7, 1.5, 1.8, \ldots$

19. $-2, -3, -5, -6, -8, -9, \ldots$

20. $8, 11, 9, 12, 10, 13, \ldots$

21. $6, 8, 9, 7, 9, 10, 8, 10, 11, \ldots$

22. $10, 8, 7, 11, 9, 8, 12, 10, 9, \ldots$

Find the figure that comes next in each sequence.

23. 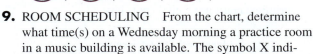 $\ldots$

24.  $\ldots$

Find the missing figure in each sequence.

25.  , , , ? ,

26. , , ? , ,

Find the next letter or letters in the sequence.

27. A, c, E, g, . . .   **28.** R, SS, TTT, . . .

29. d, h, g, k, j, n, . . .   **30.** B, N, C, N, D, . . .

What conclusion(s) can be drawn from each set of information?

31. Four people named John, Luis, Maria, and Paula have occupations as teacher, butcher, baker, and candlestick maker.

　1. John and Paula are married.

　2. The teacher plans to marry the baker in December.

　3. Luis is the baker.

　Who is the teacher?

32. In a zoo, a zebra, a tiger, a lion, and a monkey are to be placed in four cages numbered from 1 to 4, from left to right. The following decisions have been made:

　1. The lion and the tiger should not be side by side.

　2. The monkey should be in one of the end cages.

　3. The tiger is to be in cage 4.

　In which cage is the zebra?

33. A Ford, a Buick, a Dodge, and a Mercedes are parked side by side.

　1. The Ford is between the Mercedes and the Dodge.

　2. The Mercedes is not next to the Buick.

　3. The Buick is parked on the left end.

　Which car is parked on the right end?

34. Four divers at the Olympics finished first, second, third, and fourth.

　1. Diver A beat diver B

　2. Diver C placed between divers B and D.

　3. Diver B beat diver D

　In which order did they finish?

35. A green, a blue, a red, and a yellow flag are hanging on a flagpole.

　1. The blue flag is between the green and yellow flags.

　2. The red flag is next to the yellow flag.

　3. The green flag is above the red flag.

　What is the order of the flags from top to bottom?

36. Andres, Barry, and Carl each have two occupations: bootlegger, musician, painter, chauffeur, barber, and gardener. From the following facts, find the occupations of each man.

　1. The painter bought a quart of spirits from the bootlegger.

　2. The chauffeur offended the musician by laughing at his mustache.

　3. The chauffeur dated the painter's sister.

　4. Both the musician and the gardener used to go hunting with Andres.

　5. Carl beat both Barry and the painter at monopoly.

　6. Barry owes the gardener $100.

APPLICATIONS

37. JURY DUTY   The results of a jury service questionnaire are shown in Illustration 2. Determine how many of the 20,000 respondents have served on neither a criminal court nor a civil court jury.

Jury Service Questionnaire

997	Served on a criminal court jury
103	Served on a civil court jury
35	Served on both

ILLUSTRATION 2

38. ELECTRONIC POLL   In Illustration 3, the Internet poll shows that 124 people voted for the first choice, 27 people voted for the second choice, and 19 people voted for both the first and the second choice. How many people clicked the third choice, "Neither"?

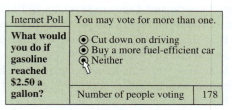

ILLUSTRATION 3

39. THE SOLAR SYSTEM   The graph in Illustration 4 shows some important characteristics of the 9 planets in our solar system. How many planets are neither rocky nor have moons?

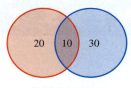

ILLUSTRATION 5

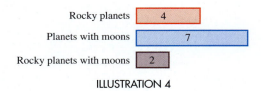

ILLUSTRATION 4

40. Write a problem in such a way that the diagram in Illustration 5 can be used to solve it.

WRITING

41. Describe deductive reasoning.

42. Describe a real-life situation in which you might use deductive reasoning.

43. Describe inductive reasoning.

44. Describe a real-life situation in which you might use inductive reasoning.

Study Set   Section 1   (page 5)

1. point, line, plane   **3.** midpoint   **5.** angle   **7.** protractor
9. right   **11.** 180°   **15.** ∠1, ∠DEF, ∠FED, ∠E   **17.** 50°
19. 25°   **21.** 40°   **23.** 130°   **25.** 180°   **27. a.** about 80°
b. about 160°   **c.** about 10°   **d.** about 100°   **29. a.** ≅
b. =   **31.** angle   **33.** ≅   **35.** 3   **37.** 3   **39.** 1   **41.** B
43. 40°   **45.** 135°   **47.** true   **49.** false; a line does not have
an endpoint   **51.** true   **53.** true   **55.** acute   **57.** obtuse
59. right   **61.** straight   **63.**

65. a. 80°   **b.** 30°   **c.** 65°

Study Set   Section 2   (page 14)

1. adjacent   **3.** supplementary   **5.** congruent   **7. a.** =
b. congruent   **c.** 33°   **d.** 2n   **9.** true   **11.** false   **13.** yes
15. yes   **17.** no   **19.** true   **21.** true   **23.** true   **25.** true
27. angle   **29.** variable   **31.** 130°   **33.** 230°   **35.** 100°
37. 40°   **39.** 10°   **41.** 27.5°   **43.** 30°, 60°, 120°
45. 25°, 115°, 65°   **47.** 53°   **49.** 35°   **51.** 60°   **53.** 75°
55. 80°   **57.** 95°   **59.** 65°, 115°   **61.** 30°

Study Set   Section 3   (page 22)

1. coplanar, noncoplanar   **3.** perpendicular   **5.** alternate,
corresponding   **11.** ≅, alternate interior   **13.** 180°, interior
15. There is not enough information to tell.   **17. a.** ∠4 and ∠6,
∠3 and ∠5   **b.** ∠1 and ∠5, ∠4 and ∠8, ∠2 and ∠6, ∠3 and ∠7
c. ∠3, ∠4, ∠5, ∠6   **19.** They are parallel.   **21.** a right angle
23. is perpendicular to   **25.** m(∠1) = 130°, m(∠2) = 50°,
m(∠3) = 50°, m(∠5) = 130°, m(∠6) = 50°, m(∠7) = 50°,
m(∠8) = 130°   **27. a.** 85°, 45°, 135°, 50°   **b.** 180°
c. 180°   **29.** 10°, 50°, 50°   **31.** 30°, 70°, 110°
33. 40°, 40°, 140°   **35.** 12°, 70°, 70°   **37.** If the stones are

level, the plumb bob string should pass through the midpoint of the
crossbar of the A-frame.   **43.** 75°, 105°, 75°

Study Set   Section 4   (page 33)

1. polygon   **3.** vertex   **5.** equilateral, isosceles, scalene
7. hypotenuse, legs   **9.** addition   **13. a.** The angles do not
have the same measure.   **b.** The sides are not the same length.
19. a. right   **b.** 90°   **c.** $\overline{AB}$, $\overline{BC}$ **d.** $\overline{AC}$ **e.** $\overline{AC}$
f. $\overline{AC}$ **21. a.** isosceles, angles   **b.** converse, length
23. a. They are the same length.   **b.** isosceles   **25.** 180°
27. triangle   **29.** segment   **31.** m($\overline{AB}$) = m($\overline{BC}$)
33. 90°   **35.** 45°   **37.** 90.7°   **39.** 55°   **41.** 50°, 50°, 60°, 70°
43. 50°, 50°, 65°, 65°   **45.** 40°, 80°, 60°   **47.** 28°   **49.** 68°
51. 12°   **53.** 39°   **55.** 40°, 70°, 70°   **57.** 39°, 102°; 70.5°,
70.5°   **59. b.** octagon   **c.** triangle   **d.** pentagon
63. equilateral

Study Set   Section 5   (page 45)

1. quadrilateral   **3.** rectangle   **5.** rhombus   **7.** diagonal
9. a. 4; A, B, C, D   **b.** 4; $\overline{AB}$, $\overline{BC}$, $\overline{CD}$, $\overline{DA}$
c. 2; $\overline{AC}$, $\overline{BD}$ **d.** yes, no, no, yes   **13. a.** 4; ∠M, ∠N,
∠O, ∠P   **b.** $\overline{MN}$ ∥ $\overline{PO}$; $\overline{NO}$ ∥ $\overline{MP}$
c. m($\overline{MN}$) = m($\overline{PO}$); m($\overline{NO}$) = m($\overline{MP}$)
d. 2; $\overline{MO}$ and $\overline{NP}$ **15. a.** 12   **b.** 6   **17.** rectangle
19. a. no   **b.** yes   **c.** no   **d.** yes   **e.** no   **f.** yes
21. a. isosceles trapezoid   **b.** ∠J, ∠M   **c.** ∠K, ∠L
d. m(∠M), m(∠L), m($\overline{ML}$)   **23. a.** 7   **b.** 5   **c.** 2
d. S = (n − 2)180°   **25.** The four sides of the quadrilateral are
the same length.   **27. a.** the sum of the measures of the angles of
a polygon; the number of sides the polygon has
b. the angle measure of a regular polygon; the number of sides the
polygon has   **c.** the measure of an exterior angle of a regular
polygon; the measure of an interior angle of a regular polygon
d. the measure of an exterior angle of a regular polygon; the
number of sides the polygon has   **29. a.** 30°   **b.** 30°   **c.** 60°
d. 8 cm   **e.** 4 cm   **31. a.** 42°   **b.** 95°   **33. a.** 9   **b.** 70°
c. 110°   **d.** 110°   **35.** 1,080°   **37.** 1,800°   **39.** 2,520°
41. 14 sides   **43.** 7 sides   **45.** 10 sides   **47.** 135°
49. 108°   **51.** 156°   **53.** 6 sides   **55.** 20 sides
57. 50 sides   **59. a.** trapezoid   **b.** square   **c.** rectangle
d. trapezoid   **e.** parallelogram   **61.** Adjust the frame so that
the diagonals are the same length.   **63.** $\frac{3}{4}$ in.; 65°; 115°

Study Set   Section 6   (page 59)

1. perimeter   **3.** area   **5.** area   **7.** 128 ft²
9. a. 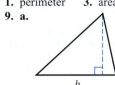   **b.**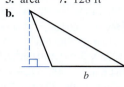
c. 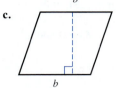   **d.**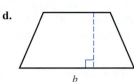

11. rectangle and triangle   **13.** length 15 in. and width 5 in.; length 16 in. and width 4 in. (answers may vary)   **15.** sides of length 5 m   **17.** base 5 yd and height 3 yd (answers may vary) **19.** length 5 ft and width 4 ft; length 20 ft and width 3 ft (answers may vary)   **21.** The length and width need to be expressed in the same units.   **23. a.** $(4x + 4)$ ft   **b.** $(4x + 6)$ ft
c. $(3b + 8)$ ft   **d.** $15x$ ft   **25.** 144   **27.** $P = 4s$
29. square inch   **31.** $A = s^2$ **33.** triangle   **35.** $A = bh$
37. 32 in.   **39.** 36 m   **41.** 3.7; 10.1; 50.8   **43.** 85 cm
45. 8.75 yd   **47.** 23.1 in.   **49.** 15 m; 10 m; 5 m; 5 m
51. 16 cm²   **53.** 60 cm²   **55.** 25 in.²   **57.** 169 mm²
59. 80 m²   **61.** 75 yd²   **63.** 75 m²   **65. a.** 12 cm
b. 30 cm   **67.** 36 ft   **69.** 189 cm²   **71. a.** 2 in.; 8 in.
b. 20 in.   **73.** $4,875   **75.** 81   **77.** linoleum   **79.** $1,200
81. $361.20   **83.** $192   **85.** 111,825 mi²   **87.** 51
89. a. 3 mi²   **b.** 11+3; 14 mi²   **c.** 8.5 mi²
d. 16; 4 mi²; 16 + 22 = 38; 9.5 mi²; 6.75 mi²

Study Set   Section 7   (page 72)

1. radius   **3.** diameter   **5.** circumference   **7.** twice
9. minor, major   **11.** arc   **13.** $\overline{OA}$, $\overline{OC}$, and $\overline{OB}$
15. $\overline{DA}$, $\overline{DC}$, and $\overline{AC}$ **17.** $\overparen{ABC}$ and $\overparen{ADC}$ **19.** $\angle COB$
21. a. Multiply the radius by 2.   **b.** Divide the diameter by 2.
23. a. 1 in.   **b.** 2 in.   **c.** 2π in. $\approx$ 6.28 in.
d. π in.² $\approx$ 3.14 in.²   **25.** Square 6.   **27.** 12 in.
29. $\overparen{YWZ}$ **31.** 318°   **33.** $\angle YXW$ **35. a.**

b. $L = \dfrac{q}{360°} \cdot 2\pi r$ **c.** $2\pi r$ **d.** $\dfrac{q}{360°}$ **37.** arc AB
39. $C = \pi D$ or $C = 2\pi r$ **41.** π **43. a.** multiplication: $2\pi r = 2 \cdot \pi \cdot r$ **b.** raising to a power and multiplication:
$\pi r^2 = \pi \cdot r^2$ **45. a.** $\dfrac{1}{4}$ **b.** $\dfrac{\pi}{2}$ **c.** $\dfrac{9\pi}{10}$ **47.** 8 in.
49. 5 cm   **51.** 37.7 in.   **53.** 36.0 m   **55.** 157.1 yd
57. 18.8 in.   **59.** 25.42 ft   **61.** 31.42 m   **63.** 7 ft
65. $\dfrac{5}{2}$ yd = 2.5 yd   **67.** 706.9 ft²   **69.** 1,963.5 in.²
71. 3.0 ft   **73. a.** 1.2 m   **b.** 2.4 m   **c.** 7.4 m   **75.** 28.3 in.²
77. 63.6 mm²   **79.** 88.3 in.²   **81.** 128.5 cm²   **83.** 27.4 in.²
85. 66.7 in.²   **87.** $\dfrac{5\pi}{6}$ in., 2.6 in.   **89.** $\dfrac{33\pi}{2}$ m, 51.8 m

91. $\dfrac{25\pi}{3}$ ft², 26.2 ft²   **93.** 375π cm², 1,178.1 cm²
95. 3.14 mi²   **97.** 32.66 ft   **99.** 12.73 times   **101.** 1.59 ft
103. 12.57 ft²; 0.79 ft²; 6.25%   **105. a.** $\dfrac{16\pi}{9}$ in. $\approx$ 5.6 in.
b. $\dfrac{17\pi}{18}$ in. $\approx$ 3.0 in.

Cumulative Review Exercises   Sections 1–7
(page 78)

1. point, line, plane   **3.** no   **4.** $\angle XYZ$, $\angle ZYX$, $\angle Y$, $\angle 1$
5. a. 110°, obtuse   **b.** 90°, right   **c.** 50°, acute
d. 180°, straight   **6. a.** measure   **b.** length   **c.** line
d. complementary   **7.** D **8. a.** false   **b.** true   **c.** true
d. true   **e.** false   **9.** 20°, 60°, 60°   **10.** 140°
11. a. transversal   **b.** $\angle 6$ **c.** $\angle 7$ **12.** m($\angle 1$) = 155°, m($\angle 3$) = 155°, m($\angle 4$) = 25°, m($\angle 5$) = 25°, m($\angle 6$) = 155°, m($\angle 7$) = 25°, m($\angle 8$) = 155°   **13.** 50°, 110°, 70°
14. Yes. Alternate interior angles are congruent if and only if the lines are parallel.   **15. a.** 8, octagon, 8   **b.** 5, pentagon, 5
c. 6, hexagon, 6   **d.** 4, quadrilateral, 4   **16. a.** isosceles
b. scalene   **c.** equilateral   **d.** isosceles   **17.** 70°
18. 70°, 70°, 40°   **19.** rectangle, square, parallelogram
21. a. 12   **b.** 13   **c.** 90°   **d.** 5   **22. a.** 10   **b.** 65°
c. 115°   **d.** 115°   **23.** 1,440°   **24. a.** 6   **b.** 60°
25. 188 in.   **26.** 15.2 m   **27.** 13 ft, 4 ft   **28.** 376 cm²
29. $(b^2 - 2b)$ square units   **30.**

31. $800   **32.** 144
33. The area is 25 ft².
34. 120 in.²   **35. a.** $\overline{CD}$, $\overline{AB}$
b. $\overline{AB}$ **c.** $\overline{OA}$, $\overline{OC}$, $\overline{OD}$, $\overline{OB}$
d. $\overparen{BD}$, $\overparen{DAB}$ or $\overparen{DCB}$ **e.** semicircle   **36.** 66.0 cm
37. 45.1 cm   **38.** 34.0 in.   **39.** 706.9 m²   **40.** 4.3 mi²
41. a. 3.5 ft   **b.** 7.0 ft   **c.** 22.0 ft   **42. a.** 47°   **b.** 313°
c. 33°   **43.** $\dfrac{11\pi}{5}$ in. $\approx$ 6.9 in.
44. $\dfrac{20,000\pi}{3}$ ft² $\approx$ 20,944.0 ft²

Study Set   Section 8   (page 89)

1. congruent   **3.** congruent   **5.** similar   **7. a.** No. They
have different sizes.   **b.** Yes. They have the same shape.
9. $\overline{DF}$, $\overline{AB}$, $\overline{EF}$, $\angle D$, $\angle B$, $\angle C$ **11.** PQR **13.** MNO
15. $\angle A \cong \angle B$, $\angle Y \cong \angle T$, $\angle Z \cong \angle R$, $\overline{YZ} \cong \overline{TR}$, $\overline{AZ} \cong \overline{BR}$, $\overline{AY} \cong \overline{BT}$,   **17.** true   **19.** False. The angles must be between the congruent sides.   **21.** true   **23.** 100   **25.** is congruent to
29. congruent   **31.** congruent   **33.** yes, SSS
35. not necessarily   **37.** yes, SSS   **39.** yes, SAS
41. yes, ASA   **43.** 12 in.; 135°   **45.** 19°; 14 m   **47.** 6 mm
49. 50°   **51.** proportional   **53.** yes   **55.** not necessarily
57. yes   **59.** not necessarily   **61.** yes   **63.** 8, 35
65. 60, 38   **67.** $\dfrac{25}{6} = 4\dfrac{1}{6}$ **69.** 16   **71.** 17.5 cm   **73.** 36 ft
75. 59.2 ft

Study Set   Section 9   (page 101)

1. hypotenuse, legs   **3.** Pythagoras   **5.** converse
7. equilateral   **9.** $a^2 + b^2 = c^2$ **11.** right   **13.** half, twice
15. $\sqrt{3}$ **17.** Subtract 25 from both sides.   **19. a.** $\overline{BC}$
b. $\overline{AB}$ **c.** $\overline{AC}$ **21.** no   **23.** hundredths

25. a. $10\sqrt{2}$ **b.** 3 **29.** $16 \cdot \sqrt{2}$ **31. a.** $b = a\sqrt{3}$
b. $c = 2a$ **c.** $c = a\sqrt{2}$ **33.** 10 ft **35.** 80 m
37. $\sqrt{74}$ cm ≈ 8.6 cm **39.** $\sqrt{51}$ in. ≈ 7.1 in.
41. $\sqrt{23}$ ft ≈ 4.8 ft **43.** 20, 21, 29 **45.** 8, 15, 17
47. $x = 2\sqrt{2} \approx 2.83, y = 2$ **49.** $x = 5\sqrt{3} \approx 8.66, y = 10$
51. $x = 4.69, y = 8.11$ **53.** $x = 12.11, y = 12.11$
55. $7\sqrt{2}$ cm **57.** $(5\sqrt{2}, 0), (0, 5\sqrt{2}), (-5\sqrt{2}, 0), (0, -5\sqrt{2})$;
$(7.07, 0), (0, 7.07), (-7.07, 0), (0, -7.07)$ **59.** $10\sqrt{3}$ mm,
17.32 mm **61.** $10\sqrt{181}$ ft, 134.54 ft **63.** about 0.13 ft
65. 5 m, 12 m, 13 m

Study Set Section 10 (page 114)

1. geometry **3.** right, oblique **5.** cube **7.** hemisphere
9. perpendicular **11. a.** circular cone **b.** sphere
c. right circular cylinder **d.** right triangular prism
e. square pyramid **f.** right regular pentagonal prism
19. $r = \dfrac{D}{2}$ or $D = 2r$ **21. a.** 24 in.^{2} **b.** 72 in.^{3}

23. 1,728 **25.** 1,000,000 **27. a.** 54 **b.** $\dfrac{500}{3}$

29. It is 2 times larger. **31.** It is 8 times larger.
33. a. cubic inch **b.** 1 cm^3 **35.** a right angle
37. 125 in.^{3} **39.** 120 cm^3 **41.** 192π ft$^3 \approx 603.19$ ft^3
43. 39π m$^3 \approx 122.52$ m^3 **45.** 60 cm^3 **47.** 48 m^3
49. 972π in.$^3 \approx 3,053.63$ in.^{3} **51.** 432π m$^3 \approx 1,357.17$ m^3
53. 100π cm$^3 \approx 314.16$ cm^3 **55.** 400 m^3 **57.** 3 ft
59. 8 yd **61.** 15 ft **63.** 30 ft^3 **65.** 41.6π m$^3 \approx 130.70$ m^3
67. $4,500\sqrt{3}$ ft$^3 \approx 7,794.23$ ft^3 **69.** 576 cm^3 **71.** 335.10 in.^{3}
73. $\frac{1}{8}$ in.$^3 = 0.125$ in.^{3} **75.** 2.125 **77.** 197.92 ft^3
79. 33,510.32 ft^3 **81.** 8:1

Study Set Section 11 (page 124)

1. surface area **3.** lateral **5.** bases **7.** right
9. ft^2, cm^2, square yards **11. a.** bases **b.** lateral
13. 180 in. **15.** $TSA = 2B + hp$; $TSA = 2\pi r^2 + 2\pi rh$;
$TSA = B + \frac{1}{2}ps$; $TSA = \pi r^2 + \pi rs$; $TSA = 4\pi r^2$

17. a. $11\sqrt{3}$ in. **b.** $121\sqrt{3}$ in.$^2 \approx 209.58$ in.^{2} **19.** 41π
21. a. 144π **b.** 72π **23.** 72 in.^{2} **25.** 5,040 in.^{2}
27. 600 in.^{2} **29.** $(48\sqrt{3} + 72)$ ft^2, 155.1 ft^2
31. $4\sqrt{3}$ cm^2, $(8\sqrt{3} + 60)$ cm^2, 73.9 cm^2 **33.** 130π ft^2,
408.4 ft^2 **35.** 315π ft^2, 989.6 ft^2 **37.** 380 yd^2 **39.** 800 yd^2
41. $225\sqrt{3}$ m; $(225\sqrt{3} + 1,800)$ m^2, 2,189.7 m^2
43. $880\pi \approx 2,764.6$ in.^{2} **45.** $224\pi \approx 703.7$ in.^{2}
47. $36\pi \approx 113.1$ in.^{2} **49.** $225\pi \approx 706.9$ in.^{2} **51.** 9 ft
53. 12 m **55.** 3 units **57.** 175π in.^{2}
59. $(432\sqrt{3} + 720)$ mm^2 **61.** 4,080 mm^2 **63.** 93,600 ft^2
65. 14 in.^{2}

Cumulative Review Exercises Sections 8–11
(page 130)

1. h. **2.** e. **3.** g. **4.** a. **5.** c. **6.** i. **7.** b. **8.** j.
9. f. **10.** d. **11.** $\angle D, \angle E, \angle F, \overline{DF}, \overline{DE}, \overline{EF}$
12. a. congruent, SSS **b.** congruent, ASA **c.** not necessarily
congruent **d.** congruent, SAS **13. a.** 8 in. **b.** 50°
14. a. yes **b.** yes **15. a.** 6 **b.** 12 **16.** 21 ft
17. a. 26 **b.** $2\sqrt{7}$ **18. a.** $2\sqrt{2} \approx 2.83, 2$
b. $\dfrac{7\sqrt{3}}{3} \approx 4.04, \dfrac{14\sqrt{3}}{3} \approx 8.08$ **c.** $9\sqrt{3} \approx 15.59, 18$
19. 31.4 in. **20.** 3 ft, 4 ft, 5 ft **21.** 1,728 in.^{3}
22. a. 125 cm^3 **b.** 480 m^3 **c.** 250π in.$^3 \approx 785.40$ in.^{3}
d. 600 in.^{3} **e.** 9,020,833.33 ft^3 **23.** 35,343 ft^3 **24.** 4 in.
25. a. 61.78 ft^2 **b.** 60π yd$^2 \approx 188.50$ yd^2 **c.** 360 mi^2
d. 100π in.$^2 \approx 314.16$ in.^{2} **26.** 24 ft

Study Set Appendix I (page A-5)

1. inductive **3.** circular **5.** alternating **7.** alternating
9. 10 A.M. **11.** 17 **13.** -17 **15.** 6 **17.** 3
19. -11 **21.** 9 **23.** **25.**
27. I **29.** m

31. Maria **33.** the Mercedes **35.** green, blue, yellow, red
37. 18,935 **39.** 0